D1708929

Developments in Soil Science 6

SOILS OF ARID REGIONS

Further Titles in this Series

1. I. VALETON
BAUXITES

2. I.A.H.R.
FUNDAMENTALS OF TRANSPORT PHENOMENA IN POROUS MEDIA

3. F.E. ALLISON
SOIL ORGANIC MATTER AND ITS ROLE IN CROP PRODUCTION

4. R.W. SIMONSON (Editor)
NON-AGRICULTURAL APPLICATIONS OF SOIL SURVEYS

5. G.H. BOLT (Editor)
SOIL CHEMISTRY (two volumes)

Developments in Soil Science 6

SOILS OF ARID REGIONS

by

H.E. DREGNE

Horn Professor and Director
International Center for
Arid and Semi-Arid Land Studies
Texas Tech University, Lubbock, Texas (U.S.A.)

ELSEVIER SCIENTIFIC PUBLISHING COMPANY
Amsterdam Oxford New York 1976

ELSEVIER SCIENTIFIC PUBLISHING COMPANY
335 JAN VAN GALENSTRAAT
P.O. BOX 211, AMSTERDAM, THE NETHERLANDS

AMERICAN ELSEVIER PUBLISHING COMPANY, INC.
52 VANDERBILT AVENUE
NEW YORK, NEW YORK 10017

ISBN 0-444-41439-8

WITH 26 ILLUSTRATIONS AND 35 TABLES

PRINTED IN THE NETHERLANDS

PREFACE

The ascendancy of man began in the arid regions of Africa and Asia. An agricultural revolution there allowed the hunting and food-gathering societies to become food-producing societies in which some members were free to pursue the arts and sciences and, in time, to form urban communities and flourishing civilizations. During the last thousand years, however, these same regions have suffered the ravages of conquest, exploitation, and neglect, to the detriment of the people and the land. Deforestation, soil deterioration, and erosion led to a decline in productivity, in population, and in cultural level. Over much of the area, the decline continues today even though effective tools and technologies are available to stem further degradation of the land.

In the New World arid lands, another independently conceived agricultural revolution made possible the development of a few advanced societies. They accomplished notable feats of engineering in the construction of irrigation works upon which the people depended for food production. Some of those civilizations had fallen into decay prior to the arrival of European conquerors; others lost their vitality after the conquest when the indigenous administrative structure was altered or destroyed. More recently, exploitation and overpopulation have had damaging effects on the stability of soil and vegetation. Some areas face a crisis in attempting to stop and reverse the trend toward further degradation of the natural environment, while others prosper from the judicious application of advanced technologies.

Belatedly, recognition is now being given to the opportunities the arid regions present and the problems they pose for hundreds of millions of people in both the developed and developing nations. Much remains to be done before the proper strategies for development of the arid lands can be enunciated and put into effective operation. A key deficiency is the dearth of knowledge about the physical and biological resources of the area. This book is intended to provide information on one physical resource — the soil — that plays a major role in determining man's ability to survive on earth.

A soil-mapping project sponsored jointly by the Food and Agriculture Organization of the United Nations and the United Nations Educational, Scientific, and Cultural Organization, with the cooperation of the International Society of Soil Science, proved to be valuable in the preparation of continental soil association maps. Detailed soil maps are not available for most of the arid regions, which made it necessary, in many instances, to infer soil conditions from geologic, physiographic, geographic, and vegetation maps. In general, the more arid the climate, the less is the accuracy of the soils information.

The first three chapters present a brief description of the arid regions; an introduction to soil classification systems, with emphasis upon the United States Comprehensive Soil Classification System; and an overview of the characteristics and distinguishing features of arid-region soils. The next six chapters describe the soils of five continents and Spain. European arid-region soils are not described in a continental context because they are inextensive except in the Soviet Union and Spain. The arid regions of the European part of the Soviet Union are an extension of arid Asia and are included in the chapter on Asia. Finally, there are three chapters on chemical, physical, and biological properties. The last three chapters are not intended to be exhaustive in either the topics covered or the detail of the discussion; the topics considered are primarily those of importance to the use and management of arid-region soils.

Although an attempt has been made to keep technical terms to a minimum, it has been impossible to avoid their use entirely. Simplified definitions of the terms most likely to be new to the nonspecialist in soil science are given in the glossary.

My deep appreciation goes to colleagues throughout the world who have generously provided me with published and unpublished soil data and maps. Without their kind assistance, my task would have been much more difficult.

H.E. DREGNE (Lubbock, Texas)

CONTENTS

Chapter 1

THE ARID REGIONS

INTRODUCTION

Soils

Arid-region soils possess many unique characteristics that distinguish them from their more well-known counterparts in the humid regions. They commonly have a low level of organic matter, slightly acid to alkaline reaction (pH) in the surface, calcium carbonate accumulation somewhere in the upper five feet of soil, weak to moderate profile development, coarse to medium texture, and low biological activity. Frequently, in both the cold and the hot arid zones, soils will be covered by a thin layer of stones and gravels that constitutes a *desert pavement*. Soluble salts may be present in quantities sufficient to influence plant growth, particularly in poorly drained depressions, in irrigated areas, and in soils containing appreciable amounts of gypsum.

The usual topographic sequence (catena) of arid-zone soils begins (Fig.1.1) with shallow rocky soils on the barren mountains and hills, then progresses downslope to coarse-textured and deeper soils on the dissected upper alluvial fans, and is followed on the lower fans and plains by finer-textured and deeper soils with more well-defined carbonate and clay horizons. The latter may or may not have sand dunes on them. Finally, at the lowest level, two quite different soil conditions may occur. If the watershed drains into a closed basin (playa), fine-textured saline or gypsiferous soils are likely to be dominant. If there is exterior drainage through a stream channel, the soils of the stream flood plain usually are of variable texture and nonsaline while those in the watercourse itself are coarse textured and nonsaline.

Understanding general soil—landscape relations is invaluable to the student of soil science and geography. Yet, it is well to remember that there are many exceptions to the general relations existing between arid-region soils and geomorphic surfaces. Texture, structure, depth, salinity, reaction and calcium carbonate content may vary markedly from that of the typical soil. The differences that are encountered can be ascribed to variations in such things as the age of the surficial materials, mineral composition, kind of vegetation, physiographic position, wind and water erosion, past climates, seasonal rainfall distribution, man's encroachments, and other factors. In combination, they are responsible for an array of soils that compares in diversity with that of the humid regions.

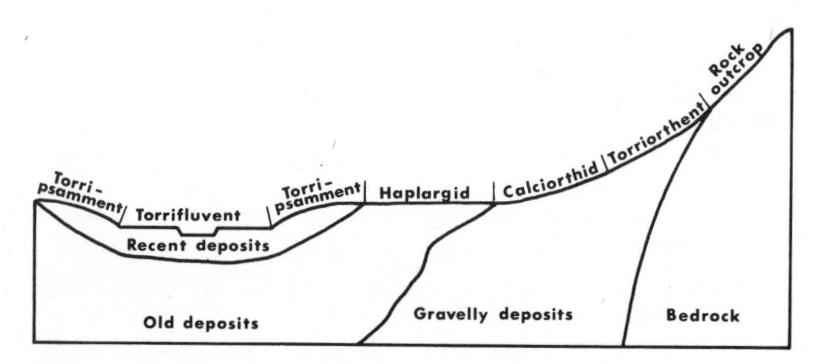

Fig.1.1. Example of soil—landscape relations in an arid region. Rock Outcrops have no significant soil cover, Torriorthents are weakly developed, Calciorthids are more developed, and Haplargids show the most development in this sequence. Torripsamments and Torrifluvents are undeveloped sands and recent alluvial soils, respectively.

Early irrigation development

Major civilizations of the past have, to a large degree, risen in those arid regions of the world where climatic and soil conditions, coupled with irrigation water, favored intensive agriculture and a settled existence. Mankind's first great cultural revolution, the development of agriculture, had its earliest origins in the semiarid Fertile Crescent of the upland plains of the Tigris, Euphrates, and Jordan river watersheds. After domestication of wheat, barley, dogs, cattle, sheep, pigs, and goats in the uplands by the hunters and food gatherers who lived there, gradual movement into the arid lowlands occurred (Braidwood, 1960). It was in the lowlands that irrigation gained ascendancy in about the year 4000 B.C. and provided the food surplusses that permitted some of the population to engage in nonagricultural pursuits. With further increases in crop production as irrigation practices improved, more and more people were freed from the necessity of growing their own food. That led to the second great cultural revolution, the growth of cities, and with it came rapid changes in society and technology (Adams, 1960). The nonfarming population became priests, tradesmen, craftsmen, soldiers, and politicians, all dependent upon a successful irrigation agriculture. With the advent of cities new institutions arose that led to larger and more complex social units than had existed before.

Mesopotamian irrigated agriculture experienced alternating periods of prosperity and adversity over thousands of years, due to wars, droughts, and floods. The final disruption came about when siltation of canals, waterlogging, and salinization of soils gradually became greater problems than the people were able to control. With deterioration of the well-developed irrigation system, the large urban communities could no longer be supported. Seven hundred years ago, Mongols invaded Mesopotamia and completed the destruction of the society that already had become weak. Today, Iraq is still trying to recover from the effects of soil destruction that occurred a millennium ago.

A different set of circumstances has permitted the Egyptians to cultivate their lands along the Nile River for centuries, without interruption. In contrast to Mesopotamia, where water distribution was tightly controlled by an elaborate system of canals that had to be maintained constantly, and annual flooding of the Tigris and Euphrates affected only a small part of the irrigated land, the Nile Valley was subjected to widespread floods each summer. Flood waters were destructive to human settlements but they served three important services: they wetted the soils thoroughly to a depth of several feet, they washed potentially damaging salts from the root zone of crops, and they deposited a thin layer of silt and clay that added plant nutrients to the fields. After the floods had subsided, fields could be planted and crops could be produced from the reservoir of water in the soil. Construction of the original Aswan Dam at the turn of the century brought the Nile under partial control; complete control was achieved when the High Aswan Dam was built. Sediment deposition on fields no longer occurs because much of the sediment settles out in Lake Nasser above the dam before the water is applied to fields. Salinization, however, now is a greater threat in the absence of periodic leaching by flood waters. Also, fertilizers must be applied in greater amounts than before to make up for the loss of sediment-borne nutrients. Neither of these unfavorable circumstances presents insurmountable problems but they do call for added expenditure and more attention to soil conditions than had been necessary in the past.

In the eastern hemisphere, other early centers of advanced cultures were established in the arid Indus Plain and in semiarid central China. Adequate irrigation water, a warm climate, rich alluvial soil, and a productive agriculture provided the favorable setting for the growth of urban communities.

Similar developments, but later and on a less-advanced level, occurred in the western hemisphere. There, the Hohokam people of the Salt River Valley in Arizona, the Aztecs of the Valley of Mexico, and the Incas of Peru utilized the same combination of natural resources to advance beyond the cultural level of the less-organized societies around them.

Other uses of soil

Arid-region soils lend themselves to uses other than that of dryland and irrigated crop production. Coarse-textured, well-drained upland soils provide, except in the very sandy areas, an excellent base for roads. Finer-textured alluvial soils serve as construction material for buildings in places where wood is a scarce commodity. Adobe bricks, made of a mixture of clay, sand, and straw, have been used for centuries to make durable structures that last for years in the arid climate if replastered periodically with mud to protect the walls against the occasional heavy rains. Mud villages are common in hot areas, not alone because soil suitable for buildings is readily available but also because thick adobe, mud, or sod walls provide good insulation during the hot days and cold nights.

Ecological equilibrium

One of the characteristics of the arid zone is that its soils, plants, and landscapes exist in a state of precarious equilibrium. The physical and biological environment is fragile, being constantly on the verge of instability and, thus, sensitive to small perturbations induced by man or nature. Recovery from a protracted drought, a single plowing of the land, or destruction of the plant or soil cover by a heavy vehicle tends to be very slow. Once the grass cover or the desert pavement has been disturbed, accelerated soil erosion may form sand dunes or arroyos that permanently alter the landscape and reduce soil productivity.

EXTENT OF THE ARID REGIONS

Scarcity of precipitation is the dominant characteristic of the arid regions of the world. Soils, vegetation, animal life, and topographic relief vary widely in the deserts of Central Asia, Australia, and North Africa, but dryness is common to all of them. Dryness, however, is not due solely to a lack of precipitation; it is influenced strongly by temperature, humidity, wind, and seasonal distribution of rain and snow. Data on mean annual precipitation, alone, do not permit delineation of the arid zones adequately even though they may be useful as first approximations (Fig.1.2). Reitan and Green (1967) have given an excellent review of the state of knowledge of the

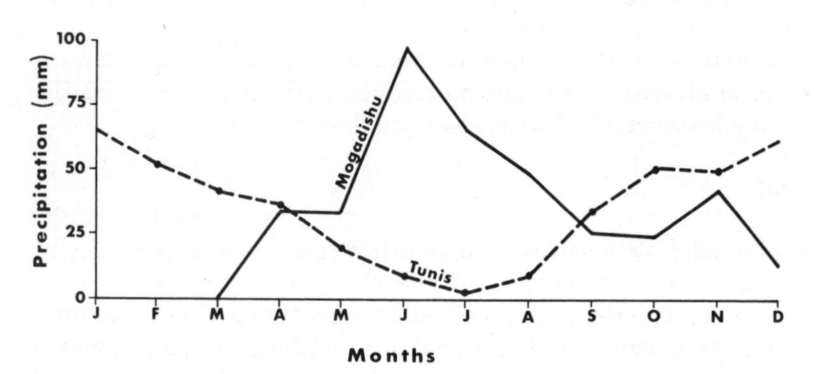

Fig.1.2. Monthly rainfall distribution in two climatic zones. Mogadishu (Somalia) (annual rainfall 426 mm, summer peak) is arid; Tunis (Tunisia) (annual rainfall 419 mm, winter peak) is semiarid.

weather and climate of arid regions and of the climatic factors controlling aridity.

The importance of climate stems from the effect of precipitation, temperature, and evaporation on the distribution of soils, vegetation, animals and geomorphic surfaces. Yet, maps of the latter four components of the environment, on a world scale, do not have boundaries that coincide with climatic boundaries. The differences are not due solely to differences in criteria. A major factor is the occurrence of relict soils (paleosols), vegetation, and land surfaces that reflect the influence of an earlier more humid climate rather than the present one. Nowhere is this shown better than in the soils of Australia and South Africa. They are among the oldest soils in the world, probably having undergone continuous development for a million or so years in alternating humid and arid climates. While some of them show characteristics associated with exposure to the present-day arid climate, others bear no resemblance to the soils of the comparatively youthful arid regions of the U.S.S.R. and North America. To further complicate the picture, man's activities have reduced or increased plant and animal life and have had locally important effects on the kinds of soils that are found.

Classification by climate

The climatic classification that currently serves best to delineate the hot arid regions of the world is that of Meigs (1953) shown in Fig.1.3. Meigs' system used the Thornthwaite (1948) moisture index for determining the portions of the earth's surface that are arid, then refined the delineation by grouping areas that have a similar seasonal distribution of precipitation and similar temperatures for the coldest and warmest months. Three degrees of aridity are identified: semiarid, arid, and extremely arid. Boundaries for the three major aridity zones were established arbitrarily by setting them so that they were in substantial agreement with generally accepted temperature limits.

On the basis of Meigs' 1953 classification, the area of semiarid, arid, and extremely arid lands is approximately as shown in Table 1.1 (Shantz, 1956). Of a total land area of 146,300,000 km^2 (56,500,000 sq. miles), about 33% is arid, excluding the cold deserts. Meigs did not attempt to include the cold arid regions in his classification.

The distribution of arid lands shown in Fig.1.3 was used to construct a diagram (Fig.1.4) of the distribution of arid lands (extremely arid, arid, and semiarid) of the world in relation to latitude. Each data point represents a transect across the map at each 10° latitude, with the width of the arid region at the latitude being expressed in kilometers. In the northern hemisphere, the bulk of the arid lands is located between 10° and 50° latitude while in the southern hemisphere it lies mainly between 20° and 30° latitude.

Péwé (1974) has estimated the glacier-free terrestrial area of polar

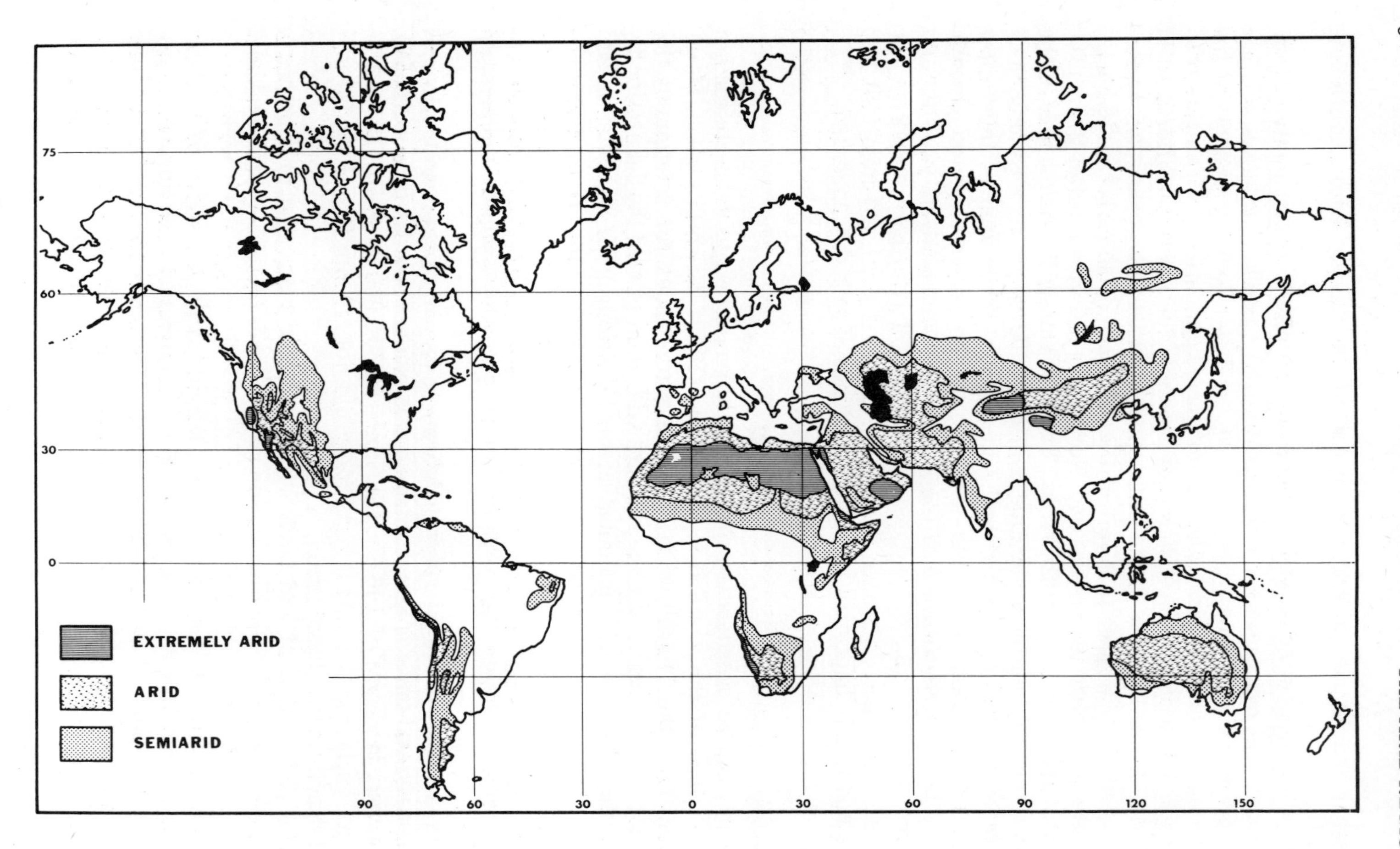

Fig. 1.3. Arid regions of the world (modified from Meigs, 1953). Polar arid regions not shown.

TABLE 1.1

Area of arid lands of the world based on climate (after Shantz, 1956)

Aridity classification		Area	
		km^2	sq. miles
Semiarid		21,243,000	8,202,000
Arid		21,803,000	8,418,000
Extremely arid		5,812,000	2,244,000
	Total	48,858,000	18,864,000

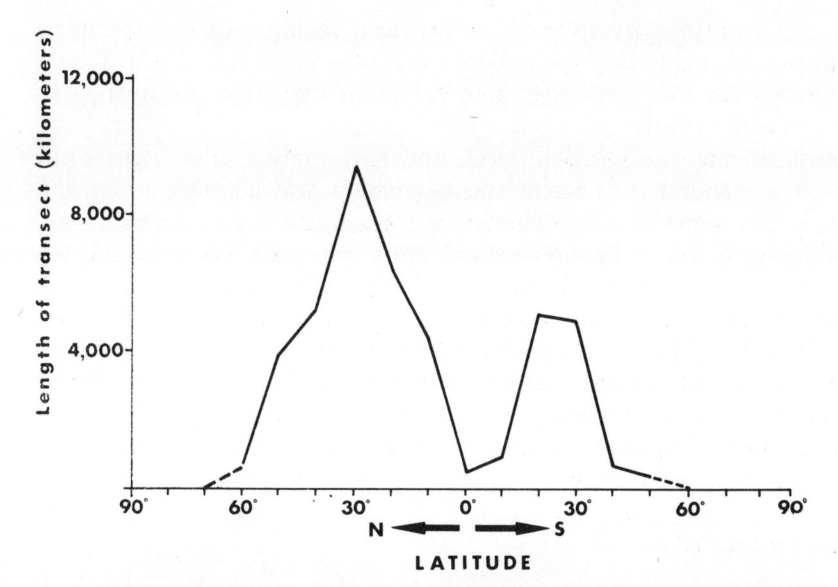

Fig.1.4. Distribution of arid lands in northern and southern hemispheres as estimated from transects along each 10° latitude. Polar arid regions not included.

deserts in the northern hemisphere to be about 4,300,000 km^2 (1,660,000 sq. miles) and in Antarctica to be about 600,000 km^2 (230,000 sq. miles). An unknown fraction of the polar deserts has soils which could be classified as arid-region soils.

Aridity is due to three general conditions occurring individually or in combination: separation by topography or distance from major moisture sources, existence of dry stable air masses, and the lack of storm systems. In the northern hemisphere, with its large land masses, distance from oceans and seas is the dominant, but not exclusive, cause of aridity. In the southern hemisphere each of the three conditions accounts for a significant portion of the area of arid lands.

Among several ways to estimate the extent of arid lands of the world, climate should be the most significant since atmospheric aridity is the causative factor for the occurrence of the condition. However, expressions of aridity as manifested by the character of the soil, vegetation, animal life, and landscape can be instructive in determining the accuracy of the climatic boundaries, particularly in places where climatic data are scarce. Distribution of animal life tends to be least useful due to the mobility of the higher life forms but the other three interrelated nonclimatic criteria (soil, vegetation, landscape) should be helpful in delineating the arid zones.

Classification by soils

The land area occupied by soils of the hot arid regions is estimated to be 31.5% of the world, excluding the polar deserts (see Table 3.2, p.39). This is somewhat more than the 28% estimated from the great soil group map of Kellogg and Orvedal (1969).

Soils do not fit into the semiarid, arid, and extremely arid categories but most Aridisols are found in the arid and extremely arid climatic zones and Mollisols are in the semiarid zone. The latter extend into the subhumid zone.

Areal delineation of cold arid-region soils has not been attempted although there is ample evidence that soils having characteristics similar to those found in hot arid regions do occur in Antarctica and elsewhere. Tedrow and Ugolini (1966) note that soils of the dry valleys of Antarctica are alkaline, may contain lime and gypsum, are usually well supplied with potentially available plant nutrients and are virtually without organic matter or biological activity. These characteristics duplicate what one would find in many soils of the hot arid regions.

Classification by vegetation

Based upon vegetation types, Shantz (1956) estimated the extent of the arid lands of the world to be that given in Table 1.2. According to these figures, about 32% of the land area is arid. This percentage agrees closely with the percentage based on climate but the proportion of land in the semiarid and arid classes is quite different in the two tables. As with Table 1.1, the cold arid regions are excluded.

Classification by drainage system

Type of regional drainage system was considered by De Martonne (1927) to be a reflection of the aridity (or lack of it) of a region. Interior-basin or closed-basin drainage systems indicate arid conditions and the consequent absence of sufficient runoff to produce rivers that would surmount topo-

TABLE 1.2

Area of the arid lands of the world based on vegetation (after Shantz, 1956)

Aridity classification		Area	
		km^2	sq. miles
Semiarid (scherophyll brush land, thorn forest, short grass)		7,045,000	2,720,000
Arid (desert grass savannah, desert grass—desert shrub)		33,411,000	12,900,000
Extremely arid (desert)		6,294,000	2,430,000
	Total	46,750,000	18,050,000

graphic barriers and provide drainage to the sea. Humid regions would be characterized by exterior- or oceanic-drainage systems. Table 1.3 gives the area of the land surface of the world that has interior drainage and is, therefore, presumed to be in an arid climate. Generally, only the very driest part of the world would have no surface runoff, whereas the semiarid regions would have occasional runoff that would result in intermittent lakes forming in depressions. Polar interior-basin drainage areas are included in the 41,838,000 km^2. According to De Martonne's estimate, 29% of the earth has the closed-basin type of drainage and 35% of the nonpolar land area is in that category.

Based on the four independent estimates, moisture limitations are of major importance over about one third of the land area of the world. Within that area aridity has played a dominant role in shaping the development of landscapes, soils, flora, and fauna and has had, and continues to have, a profound influence on man's occupancy of the region.

TABLE 1.3

Area of the arid lands of the world based on drainage system (after De Martonne, 1927)

Drainage type		Area	
		km^2	sq. miles
No surface runoff		27,991,000	10,807,000
Land surface runoff, not reaching oceans		13,847,000	5,346,000
	Total	41,838,000	16,153,000

CLIMATIC VARIABILITY

Variability in annual precipitation is characteristic of arid-zone climates (Fig.1.5). At Karachi (Naqvi, 1958) the mean annual rainfall for the 100-year period from 1850 to 1950 was 7.75 inches (197 mm) but the driest year had only 0.47 inches (12 mm) while the wettest year had 28.00 inches (710 mm), 60 times more than in the driest year. During that 100-year period two out of every three years were below the mean. Average precipitation during the below-average years was 4.15 inches (105 mm) and during the above-average years was 14.81 inches (376 mm). Within the arid regions, variability in annual precipitation increases as the total precipitation decreases. Thomas (1962) gives data on precipitation for six locations in the southwestern part of the U.S. (Table 1.4). At Indio (Calif.), the ratio of the precipitation of the wettest year to that of the driest is 35:1, whereas at Santa Fe (N.M.), with much higher total precipitation, the ratio is less than 4:1.

Monthly variations may be even more extreme than annual fluctuation. At Piedra Clavada, in San Luis province in Argentina, the maximum recorded rainfall during June was 133 mm and the minimum was 1 mm (Servicio Meterológico, 1962). The annual mean at that station is 174 mm (6.8 inches). Several Argentine stations in the arid region record no rainfall at all in some years for ten or more of the months even though the annual means range from 150 to 250 mm and the seasonal distribution is only

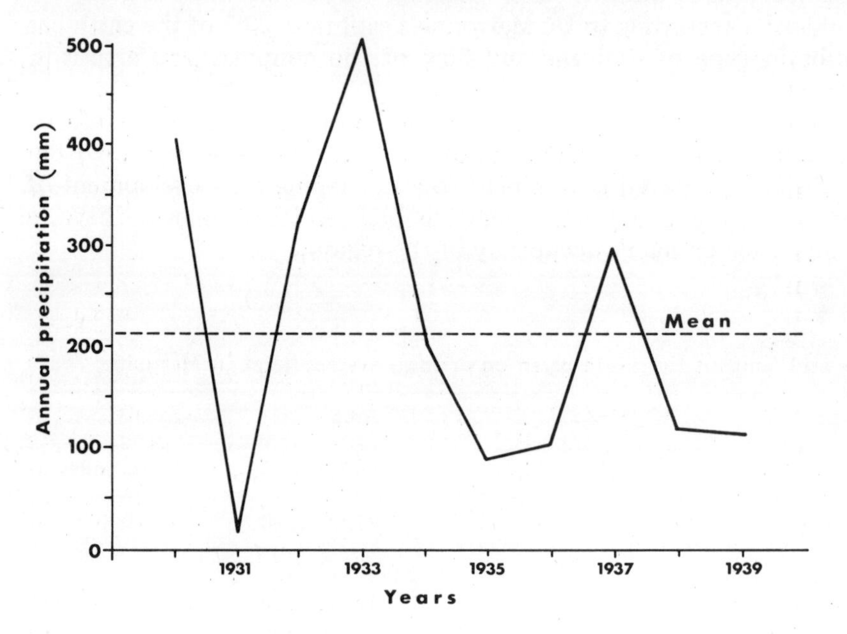

Fig.1.5. Annual rainfall variation during one ten-year period at Karachi (Pakistan).

TABLE 1.4

Annual precipitation at six U.S. weather stations

Station	Length of record (years)	Precipitation (mm)		
		max.	min.	mean
Indio (Calif.)	75	317	9	90
St. George (Utah)	63	513	76	215
El Paso (Texas)	74	443	61	220
San Diego (Calif.)	103	662	92	255
Tucson (Ariz.)	85	531	120	284
Santa Fe (N.M.)	103	680	186	358

Source: Thomas (1962).

moderately variable. In the driest sections of the world (Chile's Atacama Desert and the central Sahara) a single rain may represent the only precipitation that will fall for ten or more years. Averages are of little significance under such conditions.

Annual temperatures are essentially invariable, compared to precipitation. Over a 36-year period, annual temperatures at El Paso (Texas) varied from a low of 17.1°C (62.8°F) to a high of 19.5°C (67.1°F) (Weather Bureau, 1966). The maximum annual temperature was just 1.1 times that of the minimum. Monthly temperature variations from the mean were greater but seldom exceeded 2.5°C (4°F). Daily ranges in temperature are large, commonly amounting to 15—20°C (27—36°F) and frequently exceeding this in places where the humidity is very low. Where humidity is high, as in coastal deserts, differences in temperature extremes are much less. Relative humidity in the arid southwestern United States varies from about 60% in the early morning to 20—40% at noon (U.S. Department of Agriculture, 1941). Humidities as low as 10% or less, are not uncommon during periods when hot dry winds are blowing.

These small variations in temperature and large variations in precipitation from year to year affect soils, vegetation, and animal life differently than they affect landscape form. Soils, plants, and animals of the arid zones are a reflection of the less-than-average precipitation that falls in a majority of the years rather than of the mean annual precipitation. For example, average depth to carbonate layers in soils is a function of the large number of dry years rather than of the small number of wet years. Taylor (1934) has noted that the driest years have the most pervasive influence on vegetation and animal life. On the other hand, the minimal variation in annual temperature means that average annual temperature is a good indication of the temperature that controls chemical and biological processes in a region.

Landscape form, by contrast, is strongly affected by wet years that exert a disproportionately large influence on some components of the landscape. Even a cursory examination of arroyos and other stream channels makes it obvious that volumes of water far in excess of the average runoff must have moved through the watercourses in the past in order to account for the amount of erosion that has occurred and the size of the transported materials.

REFERENCES

Adams, R.M., 1960. The origin of cities. *Sci. Am.*, 203: 153—168.

Braidwood, R.J., 1960. The agricultural revolution. *Sci. Am.*, 203: 130—148.

De Martonne, E., 1927. Regions of interior-basin drainage. *Geogr. Rev.*, 17: 397—414.

Kellogg, C.E. and Orvedal, A.C., 1969. Potentially arable soils of the world and critical measures for their use. *Adv. Agron.*, 21: 109—170.

Meigs, P., 1953. World distribution of arid and semi-arid homoclimates. In: UNESCO, *Reviews of Research on Arid Zone Hydrology. Arid Zone Res., I:* 203—210.

Naqvi, S.N., 1958. Periodic variations in water balance in an arid region — a preliminary study of 100 years rainfall at Karachi. In: UNESCO, *Climatology and Microclimatology. Arid Zone Res., XI:* 326—345.

Péwé, T.L., 1974. Geologic and geomorphic processes of polar deserts. In: T.L. Smiley and J.H. Zumberge (Editors), *Polar Deserts.* University of Arizona Press, Tucson, Ariz., pp.33—52.

Reitan, C.H. and Green, C.R., 1967. Weather and climate of desert environments. In: W.G. McGinnies, B.J. Goldman and P. Paylore (Editors), *Deserts of the World.* University of Arizona Press, Tucson, Ariz., pp.18—92.

Servicio Meteorológico Nacional, 1962. *Datos Pluviométricos, 1921—1950.* Secretaría de Aeronáutica, República Argentina, Buenos Aires, Publicación B, No. 21.

Shantz, H.L., 1956. History and problems of arid lands development. In: G.F. White (Editor), *The Future of Arid Lands. Am. Assoc. Adv. Sci., Publ.* No. 43: 3—25.

Taylor, W.P., 1934. Significance of extreme or intermittent conditions in distribution of species and management of natural resources with a restatement of Liebig's Law of Minimum. *Ecology*, 15: 374—379.

Tedrow, J.C.F. and Ugolini, F.C., 1966. Antarctic soils. In: J.C.F. Tedrow (Editor), *Antarctic Soils and Soil Forming Processes. NAS—NRC, Publ.* No. 1418: 161—177.

Thomas, H.E., 1962. The meteorologic phenomenon of drought in the Southwest. *U.S. Geol. Surv., Prof. Pap.*, 372-A: 42 pp.

Thornthwaite, C.W., 1948. An approach toward a rational classification of climate. *Geogr. Rev.*, 38: 55—94.

U.S. Department of Agriculture, 1941. *Climate and Man. Yearbook of Agriculture, 1941.* U.S. Government Printing Office, Washington, D.C., pp.731—733.

Weather Bureau, U.S. Department of Commerce, 1966. *Local Climatological Data. 1965. El Paso, Texas.* U.S. Government Printing Office, Washington, D.C.

Chapter 2

SOIL CLASSIFICATION

INTRODUCTION

Soil classification owes much to the classical studies in Russia of V.V. Dokuchaev and his students during the last third of the 19th century. Dokuchaev, a geologist by training, began his pioneering work with a field and laboratory investigation of the characteristics of the dark-colored grassland soil (Chernozem) which stretched in an east–west band across the subhumid and semiarid central part of European Russia. After detailed study of the morphological, physical, and chemical features of vertical cross-sections (profiles) of the soils there, he noted a marked similarity in the various profiles throughout the zone despite differences in the geologic formations. His further observation that the climate was comparatively uniform throughout the area led him, in time, to formulate his hypothesis that climate was an important factor in soil development. Dokuchaev realized that the character of the parent rock or of the climate could not account for the properties of many soils, and the concept of multiple soil-forming factors (climate, topography, parent material, vegetation, and time) gradually evolved.

The role of climate in soil development was recognized independently in the United States by E.W. Hilgard, also a geologist, at about the same time when Dokuchaev made his landmark observations (Jenny, 1961). Hilgard called attention to many of the characteristics which differentiate soils of the arid regions from those of the humid regions. He noted that coarse-textured soils were predominant in the arid regions, lime accumulation in the subsoils was common, and native fertility was relatively high, all because of the lesser amount of leaching in the dry areas. Hilgard's concepts were utilized by European soil scientists in understanding soil characteristics but were largely neglected in the United States, where soil science was in its infancy.

It was no coincidence that the concept of soil as a natural body rather than as merely weathered rock arose in Russia and the United States. Both nations occupy large land masses with a variety of climates — humid and arid — and discerning scientists could see soil differences which were not present or were obscured in smaller countries where soil investigations were undertaken. From these initial observations came the realization that the distribution of soils was not haphazard but, rather, was related to the environment in which they occurred. That was a monumental step toward understanding soils, their properties, and their potential for utilization.

SOIL MAPPING

While initial emphasis of the Russian school of soil science was on how soils developed, interest in the classification and mapping of soils soon followed. In that field, the Russians were preeminent for several decades, even though the first soil surveys in the United States were started as long ago as 1899 by the U.S. Department of Agriculture in Connecticut, New Mexico, and Utah in response to requests for assistance in solving crop-production problems. The major thrust toward a comprehensive system of soil classification in the United States came with the translation into English by C.F. Marbut of K.D. Glinka's book, *Die Typen der Bodenbildung* (The Types of Soil Formation) (Marbut, 1927). Glinka was a student of Dokuchaev who went on to become famous in his own right through his work on soil classification and soil geography and the translation of his books into German and English. Marbut, who was the chief of the Division of Soil Survey of the Bureau of Chemistry and Soils in the U.S. Department of Agriculture at the time, was so impressed by the classification system presented in Glinka's book that he began to emphasize climate and vegetation as soil-forming factors in the classification systems that he was developing for the United States. He even retained some of the Russian terms (Chernozem, Podzol) for his great soil groups.

Marbut carried the climatic factor one step further than the Russians had by separating all soils into two categories: *pedalfers* and *pedocals*. Pedalfers were considered to be typical soils of the humid regions in which aluminum (Al) and iron (Fe) accumulated and lime (calcium) carbonates were absent. Pedocals were the typical soils of the arid regions, with an accumulation of lime carbonate in the profile. Useful though the concept of pedalfers and pedocals was — and still is — as a first approximation, it encountered difficulty in application to actual conditions. Not all soils of the humid regions are free of carbonates nor are accumulations of aluminum and iron absent in all soils of present-day arid regions.

The largely nonquantitative and descriptive character of the Russian and Marbut classification systems led to some confusion in their use — in the United States, at least. By the time Marbut introduced his great soil groups, soil surveys had been made throughout the country for over 30 years. In all of them, the mapping units were, and are, the series and type. A soil series was defined as "a group of soils having genetic horizons similar as to differentiating characteristics and arrangement in the soil profile, except for the texture of the surface soil, and developed from a particular type of parent material" (U.S. Department of Agriculture, 1938). Names of series usually are taken from local geographic names (e.g., Pecos series) and bear no relation to soil genesis or soil characteristics. Soil types combine the series names with the textural class of the surface soil (e.g., Pecos silt loam).

Prior to Marbut's work in the 1930's, little or no attempt was made to group soil types into higher (more general) categories. There was no reason to do so when each soil was believed to be a separate entity which merely happened to occur where it did. In Marbut's classification system, soil type was the lowest (most specific) category and, in theory, any soil type could be placed within the great soil group to which it belonged. In practice, this was frequently difficult due to the lack of laboratory data for the soils and the absence of clearly defined quantitative criteria for each great soil group. Consequently, the higher categories were seldom used except on small-scale, generalized maps of regions, states, or the nation. The gap between soil series and the higher categories remained, despite efforts to provide more quantitative definitions of those categories, until the *United States Comprehensive Soil Classification System* (henceforth also referred to as the Comprehensive System or 7th Approximation) was introduced in 1960 after going through seven revisions (approximations) (Soil Survey Staff, 1960). With some exceptions, it now is possible to assign soil series unambiguously to the proper higher categories and, for the first time, a continuum exists between soil orders and soil series.

CLASSIFICATION SYSTEMS

Soviet Union

In the Soviet Union, no complications of the kind experienced in the United States soil survey program were encountered. Almost from the beginning in the 19th century, the practice was to identify the lowest category (species, analogous to U.S. series) by the soil type (U.S. great soil group) to which it belonged. For example, a soil map might show that one of the soil species was a "leached, medium-humus, medium thick Chernozem" (Tyurin et al., 1959). All soils classified as Chernozem have that word or the adjectival form of it (Chernozemic) in the name of the mapping unit. There never is any doubt about the higher category to which the soil belongs. The main disadvantage to the U.S.S.R. species nomenclature is that the names are long and somewhat complicated (Rozov and Ivanova, 1968).

The fact that all mapping units are named in a manner which shows their relation to higher categories of classification would indicate that considerable precision in nomenclature exists in the Russian system. However, diagnostic characteristics of the main soil types of the U.S.S.R. are only semi-quantitative (Rozov and Ivanova, 1968). The system is becoming more quantitative as the need for reasonably precise limits on soil characteristics for classification purposes becomes more apparent and as appropriate soil data become available.

Australia

Among the other principal nations engaged in classification of arid-region soils, Australia has moved furthest toward the use of quantitative criteria in differentiating among soils on that continent. However, the currently used classification system is based on that of Stephens' (1962), employing great soil groups which take their names largely from Russian terminology and which are not defined quantitatively. The Australians, too, went through the stage of classifying soils on the basis of the geologic character of the parent material from which the soils were derived, although the Russian genetic system was already known. J.A. Prescott is credited with introducing the Russian concept and some of the names into the soil survey program of the nation (Stace et al., 1968).

Australia and southern Africa are unusual among the arid regions in being dominated by ancient landscapes which have not been affected by the glaciations that played such a large role in the northern hemisphere. Due to the age of the surficial materials, the influence of past humid climates is much more pronounced than is the case elsewhere. A genetic classification based on climate and vegetation, then, has less relevance in Australia than, say, in the Soviet Union where landscapes are relatively young. For that reason, soil morphology (presence, absence, and arrangement of horizons) has had much more significance than soil genesis in the classification of Australian soils. Stephens emphasized morphology more and genesis less than did Prescott, and that trend continues today. Stephens listed 47 great soil groups; Stace et al. (1968) reduced the number to 43 and grouped them according to their degree of profile development and degree of leaching. The Australian system may be described as a morphogenetic classification.

In 1960, Northcote (1971) published a classification system (key) which represented a radical departure from the previous systems and is unique among world soil classifications. Since it combines soil morphology and taxonomy, it can be said to be a morphotaxonomic classification. The key is a bifurcated one, for simplicity of organization.

Northcote's key has for its basis the concept of the *profile form.* Profile form refers to the "overall visual impact of the physical soil properties in their intimate association one with the other, and within the framework of the solum" (Northcote, 1971). The solum is the A and B horizon of a soil or, in the absence of a B horizon, the A horizon alone. Physical properties which are capable of being observed and recorded are the critical properties in the key. They are used to distinguish between soil groups at each step in the classification. The highest level of generalization is the Division (Primary Profile Form), followed by Subdivisions, Sections, Classes, and Principal Profile Forms. Letters and numbers comprise the nomenclature system.

Four divisions have been established. They are identified as O (Organic Primary Profile Form), U (Uniform Primary Profile Form), G (Gradational

Primary Profile Form), and D (Duplex Primary Profile Form). The O form consists of soils high in organic matter; the other three forms are classed according to the textural differences occurring within the solum. Soil profiles dominated by a mineral fraction having small or no textural differences are classified as Uniform. Mineral profiles having increasingly finer (more clayey) textures as the soil depth increases are called Gradational. The fourth primary profile form is the Duplex, in which the mineral fraction changes abruptly in texture from a relatively coarse-textured A horizon to a fine-textured B horizon. In the lower categories, degree of development, textural groups, structure, consistence, color, carbonate content, depth, and other factors are used to classify the soils.

France

French soil scientists have been the leaders in conducting and publishing soil surveys in the arid regions of northern Africa, centered on the Sahara. The classification system utilized by the French is morphogenetic in character. The conditions, processes, and results of the formation and evolution of soils are the basic elements of the classification (Aubert, 1968). The system is undergoing continual change as more information on development processes becomes available, and the trend is toward establishing quantitative limits for soil groups. At the highest category of classification (class), differentiation is a function of the degree of development of the soil, the mode of alteration of the minerals of the parent rock, type or distribution of organic matter, and the appearance of certain morphologic characters. Some of the names in the French system are taken from the Russian classification but several are unique, as the result of French concentration on soils of the humid tropics and Mediterranean climates, in contrast to the Soviet scientists who work largely in areas having continental climates.

FAO/UNESCO

In the course of the conduct of the Soil Map of the World project, a cooperative enterprise of the Food and Agriculture Organization of the United Nations and the United Nations Educational, Scientific, and Cultural Organization, a soil legend was developed (FAO/UNESCO, 1974). The first continental map published by the project is that for South America (FAO/UNESCO, 1971). In many respects, the classification resembles the 7th Approximation in the names of soil units and in the use of diagnostic horizons and taxonomic definitions. There are 26 major divisions in the highest category, in contrast to the 10 orders of the U.S. Comprehensive System.

U.S. COMPREHENSIVE SOIL CLASSIFICATION SYSTEM

In 1960, after several years of effort under the leadership of G.D. Smith of the U.S. Soil Conservation Service, a soil classification system was unveiled which differed radically from those in use previously. It is commonly called the "Seventh Approximation" because it had gone through six revisions before it was published under the title of *Soil Classification — A Comprehensive System — 7th Approximation* (Soil Survey Staff, 1960). Two supplements to the 1960 booklet have been published, in 1964 and 1967, and a book incorporating later revisions was awaiting publication in 1975. The book is titled *Soil Taxonomy* (Soil Survey Staff, 1976).

The U.S. Comprehensive Soil Classification System is designed for use with soil surveys. It considers soil properties that affect, as well as result from, soil genesis and properties that affect soil use. These properties include soil moisture and temperature, thickness of horizons, organic-matter content, mineralogy, exchangeable-base status, color, and other factors. Quantitative limits have been established for each property involved in the classification. Nomenclature terms have been taken mostly from Greek and Latin roots in a deliberate attempt to avoid terms already in use. Soil literature is full of words which are defined differently by different people, if they are defined at all; the Comprehensive System starts afresh. The artificiality of the nomenclature and the nearly total abandonment of previously used names brought about considerable resistance to the system, in early years, but its merits outweighed the objections, with the result that the system is now accepted widely.

The Comprehensive System is a taxonomic one in which soils are classified on the basis of their properties; genesis enters the system indirectly through the selection and definition of the differentiating properties. The earlier U.S. system (Baldwin et al., 1938) and the Russian, French, and Australian classifications are either based upon soil-genesis processes (the influence of climate and vegetation on soil development) or are a combination of genetic and taxonomic factors. Northcote's key for the classification of soils also is a taxonomic system (Northcote, 1971). The difference between a genetic and a taxonomic system is significant to the soil classifier. In a genetic system, some soils are considered to be representative of the climate and vegetation of the area in which they occur (e.g., Chernozems occur only in subhumid grasslands). They would be called *zonal* soils. Any nonrepresentative (nonzonal) soil found in a subhumid grassland would be considered an atypical soil which is too young or too old to have the properties assigned to Chernozems or which has been altered by man (cultivated). In a taxonomic system, how the properties of a soil developed or how the soil will change in the future is of no importance; soils are classified only in accordance with their measured properties.

Highest category

The highest category of classification consists of ten orders: Entisols, Vertisols, Inceptisols, Aridisols, Mollisols, Spodosols, Alfisols, Ultisols, Oxisols, and Histosols. Differentiating characteristics among the ten orders are not the same. Entisols and Inceptisols are distinguished by their minimal development and, in the case of the Inceptisols, by occurrence in a subhumid or humid environment; Vertisols by the presence of wide and deep cracks resulting from swelling and shrinking of clay; Aridisols by the presence of horizons or profile accumulations characteristic of soil development in arid regions (e.g., soluble salts, carbonates); Mollisols by a relatively dark A horizon and more than 50% base saturation in the deeper horizons; Spodosols by a horizon of accumulation of amorphous humus, iron, and aluminum; Alfisols by being well leached and having a horizon of accumulation of crystalline aluminosilicate clay accompanied by more than 35% base saturation in the deeper layers; Ultisols by being well leached, having a horizon of clay accumulation (dominantly kaolinite), and possessing a base saturation of less than 35% in the deeper layers; Oxisols by containing highly weathered residual concentrations of free oxides (e.g., iron and aluminum oxides) and kaolinitic clays; and Histosols by having large amounts of organic matter typical of peats and mucks and by being saturated with water for most of the year.

Although soil genesis, as related to climate and vegetation, is not a formal part of the Comprehensive System, it has been a factor in selection of differentiating properties for the orders. Aridisols are soils of the arid shrublands, Mollisols of semiarid and subhumid grasslands, Alfisols of the cool forests and the hot savannahs, Spodosols of the cold and hot forests; Ultisols of the warm forests; and Inceptisols of the subhumid to humid grasslands and forests. Entisols, Vertisols, Oxisols, and Histosols are not limited to particular climates and vegetation even though Vertisols occur almost exclusively in arid to subhumid climates, Oxisols principally in the hot and wet tropics, and Histosols most extensively in cool or cold regions. The conscious or unconscious genetic bias of many soil scientists leads them to associate certain orders with certain climates even when the relation is a tenuous one.

Two soil properties heavily dependent upon climate are major components of the classification system: soil moisture regime and soil temperature regime. While there is generally a fairly close correlation between the climate of the atmosphere and the moisture and temperature of the adjoining soil, enough differences do occur to make it necessary to use soil data in the classifying process. Doing so serves to confine all the variables of the system to measured soil properties rather than permitting some of them to be external to the soil itself. Soil moisture and temperature regimes enter the classification at the suborder or lower level.

NOMENCLATURE

At first glance, the U.S. Comprehensive Soil Classification System appears to be complicated and cumbersome because it utilizes many newly coined words, parts of which are combined to form other words. In practice, however, the system is reasonably easy to use and the information content is high. The FAO/UNESCO Soil Map of the World project has adopted much of the terminology of the U.S. Comprehensive System.

Order

All of the terms are intended to be connotative although there are a few nonsense syllables included where no satisfactory Greek or Latin root was available. Aridisols are soils of the arid regions and the name connotes dryness. Names of suborders, great groups, subgroups, and families, in descending order of categorical hierarchy, of the Aridisols will have the formative element *id* as the final syllable in the name. *Id* is taken from the last half of *arid*. Similarly, any name ending in *oll* indicates that the soil is a Mollisol. Names of the orders, the formative element used in the names of suborders, great groups, and subgroups, the derivation of the formative element, and a mnemonicon which serves as both a memory aid and a pronunciation aid are given in Table 2.1.

TABLE 2.1

Formative elements of soil-order names

Name of order	Formative element	Derivation of formative element	Mnemonicon
Entisol	ent	nonsense syllable	recent
Vertisol	ert	L. *vertere*, to turn	invert
Inceptisol	ept	L. *inceptum*, beginning	inception
Aridisol	id	L. *aridus*, dry	arid
Mollisol	oll	L. *mollis*, soft	mollify
Spodosol	od	Gr. *spodos*, wood ash	Podzol; odd
Alfisol	alf	nonsense syllable	Pedalfer
Ultisol	ult	L. *ultimus*, last	ultimate
Oxisol	ox	F. *oxide*, oxide	oxide
Histosol	ist	Gr. *histos*, tissue	histology

Source: Soil Survey Staff (1960).

Suborder

Construction of suborder names begins with the formative element (*id* for Aridisols) which identifies the order, then attaches a connotative prefix. The Argid suborder is an Aridisol with an argillic (clayey) diagnostic horizon, the *arg* coming from *argillic*, which is derived from the Latin word argilla (clay). Formative elements for suborder names, their derivation, mnemonicons, and connotations are shown in Table 2.2.

Great group

Great groups have one or more connotative prefixes attached to the suborder name. A Natrargid is an Argid with a natric (sodic) horizon, natric being derived from natrium (sodium). Durargids are Argids with a cemented layer called a *duripan*, from the Latin *durus* (hard). Some great-group names have two connotative elements in front of the suborder name: Nadurargids are Argids with both a natric and a *dur*ipan horizon. Table 2.3 lists the formative elements for great-group names, with derivation, mnemonicon, and connotation for each element.

Subgroup

Subgroup names consist of one or more connotative adjectives placed before the great-group name. A Typic Natrargid is a Natrargid whose properties are typical of the central concept of the Natrargid great group of soils. In addition to the typic subgroup, there also are intergrade and extragrade subgroups. The intergrade subgroup has some properties of a different order, suborder, or great group than the great group in which it has been placed. A Borollic Natrargid is an intergrade subgroup because it describes a Natrargid having some properties resembling those of the Boroll suborder. Extragrade subgroups have unusual properties which are different from those of the typical great group but do not conform to those of any known soil. A common extragrade subgroup in arid regions is a lithic subgroup such as the Lithic Torriorthent, an Entisol showing a gradual decrease in organic matter with depth and occurring in a dry region (Torriorthent), with hard rock underlying the soil at a depth of less than 50 cm (Lithic).

Family

The next lower category, the family, contains the name of the subgroup followed by one or more adjectives specifying properties relevant to the use and management of soils. A Typic Natrargid, fine, montmorillonitic, thermic is a typic Natrargid having a fine (clayey) texture, containing dominantly montmorillonitic clays, and having a thermic (hot) temperature regime.

TABLE 2.2

Formative elements of suborder names

Formative element	Derivation	Mnemonicon	Meaning or connotation
alb	L. *albus*, white	albino	presence of albic horizon (a bleached eluvial horizon)
and	modified from Ando	Ando	Ando-like
aqu	L. *aqua*, water	aquarium	characteristics associated with wetness
ar	L. *arare*, to plow	arable	mixed horizons
arg	modified from argillic horizon; L. *argilla*, white clay	argillite	presence of argillic horizon (a horizon with illuvial clay)
bor	Gr. *boreas*, northern	boreal	cool
ferr	L. *ferrum*, iron	ferruginous	presence of iron
fibr	L. *fibra*, fiber	fibrous	least-decomposed stage
fluv	L. *fluvius*, river	fluvial	flood plains
hem	Gr. *hemi*, half	hemisphere	intermediate state of decomposition
hum	L. *humus*, earth	humus	presence of organic matter
lept	Gr. *leptos*, thin	leptometer	thin horizon
ochr	Gr. *ochros*, pale	ocher	presence of ochric epipedon (a light-colored surface)
orth	Gr. *orthos*, true	orthophonic	the common ones
plag	modified from Ger. *Plaggen*, sod		presence of plaggen epipedon
psamm	Gr. *psammos*, sand	psammite	sand textures
rend	modified from Rendzina	Rendzina	Rendzina-like
sapr	Gr. *sapros*, rotten	saprophyte	most-decomposed stage
torr	L. *torridus*, hot, dry	torrid	usually dry
trop	modified from Gr. *tropikos*, of the solstice	tropical	continually warm
ud	L. *udus*, humid	udometer	of humid climates
umbr	L. *umbra*, shade	umbrella	presence of umbric epipedon (a dark-colored surface)
ust	L. *ustus*, burnt	combustion	of dry climates, usually hot in summer
xer	Gr. *xeros*, dry	xerophyte	annual dry season

Source: Buol et al. (1973).

Soil-family adjectives identify soil properties useful in making interpretations of soil surveys. In the sequence in which they follow the subgroup names, the properties identified on the family name are textural class, mineralogy class, calcareous and reaction (pH) classes, soil-temperature class, soil-depth class, soil-slope class, soil-consistence class, coatings on soil particles, and the class of permanent cracks. Usually, only one to four property classes are

TABLE 2.3

Formative elements of great-group names

Formative element	Derivation	Mnemonicon	Meaning or connotation
acr	modified from Gr. *akros*, at the end	acrolith	extreme weathering
agr	L. *ager*, field	agriculture	an agric horizon
alb	L. *albus*, white	albino	an albic horizon
and	modified from Ando	Ando	Ando-like
anthr	Gr. *anthropos*, man	anthropology	an anthropic epipedon
aqu	L. *aqua*, water	aquarium	characteristic associated with wetness
arg	modified from argillic horizon; L. *argilla*, white clay	argillite	an argillic horizon
calc	L. *calx*, lime	calcium	a calcic horizon
camb	L. *cambiare*, to exchange	change	a cambic horizon
chrom	Gr. *chroma*, color	chroma	high chroma
cry	Gr. *kryos*, coldness	crystal	cold
dur	L. *durus*, hard	durable	a duripan
dystr	modified from Gr. *dys*, ill; dystrophic, infertile	dystrophic	low base saturation
eutr, eu	modified from Gr. *eu*, good; eutrophic, fertile	eutrophic	high base saturation
ferr	L. *ferrum*, iron	ferric	presence of iron
frag	modified from L. *fragilis*, brittle	fragile	presence of fragipan
fragloss	compound of fra(g) and gloss		see the formative elements frag and gloss
gibbs	modified from gibbsite	gibbsite	presence of gibbsite
gloss	Gr. *glossa*, tongue	glossary	tongued
hal	Gr. *hals*, salt	halophyte	salty
hapl	Gr. *haplous*, simple	haploid	minimum horizon
hum	L. *humus*, earth	humus	presence of humus
hydr	Gr. *hydôr*, water	hydrophobia	presence of water
hyp	Gr. *hypnon*, moss	hypnum	presence of hypnum moss
luo, lu	Gr. *louo*, to wash	ablution	Illuvial
moll	L. *mollis*, soft	mollify	presence of mollic epipedon
nadur	compound of na(tr) and dur		
natr	modified from natrium, sodium		presence of natric horizon
ochr	Gr. *ochros*, pale	ocher	presence of ochric epipedon (a light-colored surface)
pale	Gr. *palaios*, old	paleosol	old development
pell	Gr. *pellos*, dusky		low chroma
plac	Gr. *plax*, flat stone		presence of a thin pan

TABLE 2.3 (*continued*)

Formative element	Derivation	Mnemonicon	Meaning or connotation
plag	modified from Ger. *Plaggen*, sod		presence of plaggen horizon
plinth	Gr. *plinthos*, brick		presence of plinthite
quartz	Ger. *quarz*, quartz	quartz	high quartz content
rend	modified from Rendzina	Rendzina	Rendzina-like
rhod	Gr. *rhodon*, rose	rhododendron	dark-red colors
sal	L. *sal*, salt	saline	presence of salic horizon
sider	Gr. *sideros*, iron	siderite	presence of free iron oxides
sombr	Fr. *sombre*, dark	somber	a dark horizon
sphagno	Gr. *sphagnos*, bog	sphagnum moss	presence of sphagnum moss
torr	L. *torridus*, hot and dry	torrid	usually dry
trop	modified from Gr. *tropikos*, of the solstice	tropical	continually warm
ud	L. *udus*, humid	udometer	of humid climates
umbr	L. *umbra*, shade	umbrella	presence of umbric epipedon
ust	L. *ustus*, burnt	combustion	dry climate, usually hot in summer
verm	L. *vermis*, worm	vermiform	wormy, or mixed by animals
vitr	L. *vitrum*, glass	vitreous	presence of glass
xer	Gr. *xeros*, dry	xerophyte	annual dry season

Source: Buol et al. (1973).

shown in the family name. Definitions of the classes are given in the various publications of the U.S. Soil Conservation Service dealing with the Comprehensive System.

Series

Series is the lowest category. At the series level, the U.S. Comprehensive System undergoes an abrupt change in terminology: the soil-series name has no connotative value. It continues to be taken, usually, from a local name in the area where the series was described first. The Pecos series, for example, was named in 1899 when the first soil survey was initiated in the arid Pecos Valley of New Mexico and Texas. The practice of naming soil series for a local geographic feature has continued since that time. The absence of connotative terms for soil series means that the name gives no indication of how a particular series differs from another in the same family. Individual series descriptions must be consulted to obtain that information. If connotative terms were added to the family name, the series name would be so long that it would be cumbersome to use.

MOISTURE REGIMES

Suborder, great group, and subgroup names frequently include formative elements (aqu, torr, ust, ud, xer) or adjectives (aquic, aridic or torric, ustic, udic, xeric) indicating soil moisture regimes. The moisture limits for the various regimes are presented in Table 2.4. As can be seen, the limits are more qualitative than quantitative; there is no general agreement about the definition of terms such as *available* water or about the significance of variable moisture regimes during the year. Soil moisture regime is the prime criterion for the Aridisol order: an Aridisol must have an aridic soil moisture regime. Other orders may include several soil moisture regimes.

TABLE 2.4

Soil moisture regimes

Regime	Characteristics[1,2]
Aquic (aqu)	soil saturated in part or in whole long enough to remove virtually all dissolved oxygen
Aridic and torric (torr)	(1) dry in all parts of the moisture control section more than half the time that the soil temperature at 50 cm is > 5° C and (2) no period as long as 90 consecutive days when soil is moist in some or all parts of the moisture control sections while soil temperature at 50 cm is > 8° C
Ustic (ust)	soil dry in some or all parts of the moisture control section for more than 90 cumulative days in most years; soil moist in some part of the moisture control section for more than 180 cumulative days or continuously moist in some part for at least 90 cumulative days
Udic (ud)	soil not dry in any part of the moisture control section for as long as 90 cumulative days in most years
Xeric (xer)	soil dry in all parts of the moisture control section for 45 or more consecutive days within the 4 months following the summer solstice, in most years (Mediterranean climate: wet winters, dry summers)

[1] General guidelines. See *Soil Taxonomy* for definition of moisture control section and details of regime definitions.
[2] *Dry* means soil moisture tension of 15 bar or more. *Moist* means soil moisture tension greater than zero but less than 15 bar.

TEMPERATURE CLASSES

Soil-temperature classes are of special interest because knowledge of soil temperature permits several inferences to be made about the chemical and biological characteristics of the soil and, in some cases, of physical characteristics. Soil-temperature classes employed in the U.S. Comprehensive System are given in Table 2.5. Where data for actual soil temperatures at 50 cm (20 inches) are not available, a reasonably satisfactory approximation can be made by adding 1°C to the average annual air temperature.

INTERRELATIONS OF CRITERIA

Putting the U.S. Comprehensive Soil Classification System into use requires knowledge of the limits which have been established for the many criteria employed in the system. In several instances, the criteria vary, depending upon other properties of the soil. A mollic horizon is defined in part as a dark-colored horizon but not all dark-colored horizons qualify as mollic. Recourse must be made to descriptions found in *Soil Taxonomy* or other publications for details of the criteria and their application.

TABLE 2.5

Soil temperature regimes[1]

Class	Seasonal temperature difference[2]		Mean annual soil temperature (MAST)	
	°C	°F	°C	°F
Pergelic	—	—	<0	<32
Cryic	—	—	0–8	32–47
Frigid[3]	>5	>9	<8	<47
Mesic	>5	>9	8–15	47–59
Thermic	>5	>9	15–22	59–72
Hyperthermic	>5	>9	>22	>72
Iso (thermic, mesic, etc.)	<5	<9		

[1] Soil temperature at 50 cm (20 inches).
[2] Difference between average temperature in winter and summer.
[3] Frigid regime has higher summer temperatures than the cryic regime.

DEFINITIONS

Certain terms having specialized meanings are used widely in soil literature. Among these terms are: horizon, pedon, epipedon, and solum. Their relation to the soil profile is diagrammed in Fig.2.1.

A soil *horizon* is a layer which is more or less parallel to the soil surface. It differs in some property or properties from layers above and below it. The difference may be in texture, structure, color, mineralogy, or a host of other properties. Sometimes the difference is of little significance to the use and management of the soil, other times it may be of great significance. In the U.S. Comprehensive Soil Classification System, those horizons which are considered to be significant to the classification of a soil are called diagnostic horizons. They may be surface or subsurface horizons.

A soil *pedon* is a block of soil having a surface area ranging from about 1 m^2 to 10 m^2 and a variable depth extending down to the underlying geologic material which has not been altered significantly by soil-forming processes. It is a vertical cross-section of a soil, containing the horizons and layers which represent the variabilities in properties of the soil. Sampling a pedon for analysis or observation should provide material which will permit the properties of the soil to be described accurately and completely. Since a pedon is supposed to be a representative subsection of a soil, its minimum dimensions can best be determined from a roadside cut or a trench which will show the extent to which variations occur.

The *epipedon* is a soil horizon formed at the surface of the soil. It has

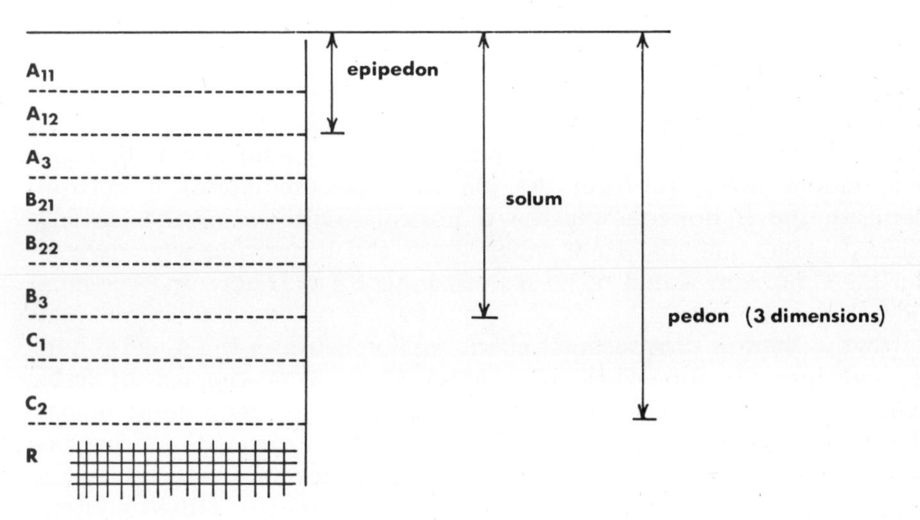

Fig.2.1. Diagrammatic representation of relation of epipedon, solum, and pedon to soil horizons.

been darkened by organic-matter additions or shows some degree of weathering or loss of soil material by leaching. In many cases, the epipedon is the same as the upper part of the A horizon of a soil but this is not true in all cases. The epipedon includes part or all of the underlying B horizon if the darkening by organic matter extends down that far.

The *solum* is the upper part of the soil profile. It shows evidence of having been altered by soil-forming processes. In a soil having only A and C horizons, the solum is the A horizon. For soils having A, B, and C horizons, it is the A and B horizons together. Northcote (1971) defines the solum in this way and states its importance in his classification system by noting that "the solum is the object which the factual key is designed to classify". In the U.S. Comprehensive System, the solum is the A and B horizons plus horizons of fragipans and duripans.

HORIZON DESIGNATION

As a carry over from the days when there was a strong genetic bias among soil classifiers, horizons and layers frequently are designated as O, A, B, C, or R. In the classic case of a soil which was developed from the weathering in place of consolidated bedrock material such as granite or limestone, the O, A, B, and C horizons are the vertical sequence of horizons from the organic matter on top of the soil down to the bedrock (the R horizon). The O horizon is a layer of undecomposed or partially decomposed plant and animal remains (organic matter) lying on top of the mineral (inorganic) soil. The litter found on top of the soil in pine forests is an O horizon. Cultivated soils do not have O horizons because the litter has been incorporated into the underlying soil.

Historically, the A—B—C sequence of horizons has been utilized to reflect presumed soil-genesis processes. The A horizon was considered to be an eluviated (leached) surface horizon from which clay, humus, and other substances moved down to form the illuviated (accumulation) B horizon. Underneath the B horizon was the C horizon, representing the partially weathered parent material from which the A and B horizons were derived. Under the C horizon would be an R horizon (U.S.) or D horizon (Australia) if bedrock were present.

Currently, there is disagreement about what constitutes the A and B horizons, and there is somewhat of a trend away from the use of letter designations for horizons because of the ambiguity in definitions. In particular, there are differing opinions about the B horizon. Some persons believe that the only true B horizon is an illuviated one in which clay has accumulated as the result of leaching from upper horizons. Others contend that a B horizon only needs to have more clay than the horizons above or below it, irrespective of whether the clay has accumulated by leaching from upper horizons or by weathering of minerals in place to form residual clays. Still others believe that a B horizon simply is one which differs in some property or properties from horizons above and below it.

In the United States, the O—A—B—C—R horizon designation is used routinely to name horizons but it is not a part of the Comprehensive System. Instead, diagnostic horizons (mollic, umbric, etc.) are employed. The diagnostic horizons may or may not coincide with the "genetic" horizons. In the Soviet Union, Australia, and France, the A—B—C terminology is an integral part of the classification systems.

In the letter system of horizon designations, subdivisions of horizons are given subscript numbers. An A_1 horizon is a subdivision of the A horizon, and A_{11} and A_{12} horizons are subdivisions of the A_1 horizon. Similarly, B_{21} and B_{22} horizons are subdivisions of the B_2 horizon which, in turn, is a subdivision of the B horizon. Lower-case letters may also be attached as subscripts: B_{21t} is a B_{21} horizon containing illuvial clay (t). *Soil Taxonomy* and the *Soil Survey Manual* (Soil Survey Staff, 1976 and 1951 resp.) may be consulted for definitions of horizon subdivisions and the letter symbols, as well as for other details of horizon nomenclature.

DIAGNOSTIC HORIZONS

Certain soil horizons are called diagnostic horizons in the Comprehensive System and they play a major role in the classification of the higher categories. Diagnostic horizons are divided into two kinds: surface horizons (epipedons) and subsurface horizons.

There are six surface and numerous subsurface diagnostic horizons. Some of them are listed below with brief descriptions. For detailed definitions — too lengthy to be included here — reference must be made to *Soil Taxonomy* or other publications dealing with the *7th Approximation*.

Diagnostic horizons formed at soil surface

Mollic epipedon: a thick (25 cm), dark horizon or horizons having a base saturation of more than 50%, with calcium the main extractable metallic cation. The structure cannot be massive or single-grained.

Anthropic epipedon: a mollic-like horizon or horizons containing more than 250 ppm of citric acid soluble P_2O_5.

Umbric epipedon: a dark horizon or horizons not having all the properties of a mollic or anthropic horizon.

Plaggen epipedon: a thick (50 cm) horizon formed by long-continued manure applications.

Histic epipedon: a thin (30 cm) organic layer of peat or muck which is normally saturated with water.

Ochric epipedon: a horizon which does not meet the criteria for the other diagnostic surface horizons.

Diagnostic horizons formed below soil surface

Among the subsoil diagnostic horizons, the following are found in arid-region soils or in soils associated with them:

Agric horizon: a horizon of accumulation of illuvial clay and humus formed under cultivation.

Argillic horizon: a horizon of accumulation of illuvial silicate clay.

Calcic horizon: a horizon of accumulation of appreciable amounts of calcium carbonate.

Cambic horizon: a horizon of alteration, with structural development and without evidence of illuviation.

Duripan: a horizon indurated with silica sufficiently to prevent slaking of dry fragments.

Fragipan: a horizon that is not indurated but which restricts movement of water and penetration of roots.

Gypsic horizon: a horizon of accumulation of appreciable amounts of gypsum (hydrated calcium sulfate).

Natric horizon: an argillic horizon having appreciable amounts of exchangeable sodium and prismatic or columnar structure.

Oxic horizon: a horizon having a residual concentration of 1:1 lattice clays and free sesquioxides with very low cation-exchange capacity.

Petrocalcic horizon: an indurated calcic horizon.

Petrogypsic horizon: an indurated gypsic horizon.

Plinthite: a horizon containing little humus but high in sesquioxides which harden irreversibly upon repeated wetting and drying.

Salic horizon: a horizon of accumulation of appreciable amounts of salts more soluble than gypsum.

Diagnostic horizons are not mutually exclusive entities. A mollic epipedon may contain an argillic subsurface horizon and plinthite may be found in ochric epipedons or in argillic horizons. This means that care must be taken in not overlooking other properties of horizons after the principal horizon type has been identified.

EQUIVALENT TERMS

Categories used in the U.S. Comprehensive Soil Classification System do not correspond in entirety to categories employed in earlier United States systems or to systems developed in other countries. Prior to the introduction of the 7th Approximation, the revised genetic classification system described in the 1938 Yearbook of Agriculture (Baldwin et al., 1938) was the accepted system in the United States. Great soil groups were the heart of the system; most of them were the same as those named by Marbut previously. Approximate equivalents between U.S. great soil groups and the orders of the Comprehensive System are presented in Table 2.6. Tables 2.7, 2.8, 2.9 and 2.10 provide the same information for the appropriate U.S.S.R., Australian, French, and FAO/UNESCO categories and U.S. soil orders.

TABLE 2.6

Approximate equivalents between U.S. great soil groups and U.S. Comprehensive System soil orders

Great soil groups	Orders in new classification
Alluvial soils	Entisols
Brown soils	Aridisols, Mollisols
Brown Forest soils	Inceptisols
Chernozem	Mollisols
Chestnut soils	Mollisols
Dark Gray and Black soils of Subtropics and Tropics	Vertisols
Degraded Chernozem	Alfisols, Mollisols
Desert soils	Aridisols
Gray-Brown Podzolic soils	Alfisols (mainly suborder Udalfs)
Gray Wooded soils	Alfisols
Hydromorphic soils	(Aquic taxa of various orders)
Latosols	Oxisols, Ultisols, Inceptisols
Lithosols	(Lithic subgroups of several orders)
Organic soils	Histosols
Podzols	Spodosols
Prairie soils	Mollisols (mainly suborder Udolls)
Reddish Brown soils	Aridisols, Alfisols
Red Desert soils	Aridisols
Reddish Chestnut soils	Mollisols, Alfisols
Red-Yellow Mediterranean soils	Alfisols (mainly suborder Xeralfs)
Red-Yellow Podzolic soils	Ultisols
Regosols	Entisols (mainly suborder Psamments)
Rendzina	Mollisols (suborder Rendolls)
Saline soils	Aridisols
Serozem	Aridisols
Terra Rossa	Alfisols (mainly suborder Xeralfs)
Tundra	Inceptisols
Weakly podzolized soils	Spodosols, Inceptisols, Alfisols

Source: Kellogg and Orvedal (1969).

TABLE 2.7

Approximate correlation between U.S.S.R. soil types and U.S. Comprehensive System soil orders

U.S.S.R.	U.S. Comprehensive System
Polar Belt	
Arctic soils	Aridisols
Tundra soils	Inceptisols
Boreal Belt	
Taiga	Inceptisols, Spodosols
Podzol	Spodosols
Volcanic	Inceptisols
Subboreal Belt	
Chernozems	Mollisols
Gray Forest	Alfisols
Chestnut	Mollisols
Desertic	Aridisols, Entisols
Serozems	Aridisols
Subtropical and Moderately Warm Belts	
Cinnamonic	Alfisols
Jeltozems (Zheltozems)	Alfisols
Kraznozems	Alfisols, Ultisols
Tropical and Equatorial Belts	
Black tropical soils	Vertisols
Desert red tropical soils	Alfisols, Ultisols
Ferralsols	Ultisols, Oxisols
Miscellaneous	
Solonetz	Aridisols, Mollisols
Solonchak	Aridisols
Hydromorphic soils	(several orders)
Sands	Entisol
Mountainous soils	(Lithic subgroups of several orders)

TABLE 2.8

Approximate correlation between Australian great soil groups and U.S. Comprehensive System soil orders

Australia	U.S. Comprehensive System
No profile differentiation	
Solonchaks	(several orders)
Alluvial soils	Entisols
Lithosols	(several orders)
Calcareous sands	Entisols
Siliceous sands	Entisols, Aridisols
Earthy sands	Entisols

TABLE 2.8 (*continued*)

Australia	U.S. Comprehensive System
Minimal profile development	
Grey-brown and red calcareous soils	Aridisols, Entisols
Desert loams	Aridisols
Red and brown hardpan soils	Aridisols
Grey, brown, and red clays	Vertisols
Dark soils	
Black earths	Vertisols
Rendzinas	Mollisols
Chernozems	Mollisols
Prairie soils	Mollisols, Alfisols
Wiesenboden	(several orders)
Mildly leached soils	
Solonetz	Mollisols, Alfisols
Solodized solonetz and solodic soils	Mollisols, Alfisols
Soloths	Alfisols, Mollisols
Solonized brown soils	Alfisols, Aridisols
Red-brown earths	Alfisols, Aridisols
Non-calcic brown soils	Alfisols
Chocolate soils	Mollisols
Brown earths	Mollisols
Soils with predominantly sesquioxidic clay minerals	
Calcareous red earths	Aridisols
Red earth	Aridisols, Alfisols
Yellow earths	Alfisols
Terra rossa soils	Alfisols
Euchrozems	Alfisols
Xanthozems	Alfisols
Krasnozems	Alfisols, Ultisols
Mildly to strongly acid and highly differentiated	
Grey-brown podzolic soils	Alfisols
Red podzolic soils	Spodosols
Yellow podzolic soils	Spodosols
Brown podzolic soils	Spodosols
Lateritic podzolic soils	Spodosols
Gleyed podzolic soils	Spodosols
Podzols	Spodosols
Humus podzols	Spodosols
Peaty podzols	Spodosols, Histosols
Dominated by organic matter	
Alpine humus soils	Spodosols
Humic gleys	Spodosols
Neutral to alkaline peats	Histosols
Acid peats	Histosols

TABLE 2.9

Approximate correlation between French soil classes and U.S. Comprehensive System soil orders

French	U.S. Comprehensive System
Sols minéraux bruts	Entisols
Sols peu évolués	Entisols, Inceptisols
Sols calcomagnésimorphes	Mollisols, Aridisols
Vertisols et paravertisols	Vertisols
Sols isohumiques	Mollisols, Aridisols
Sols à mull	(several orders)
Podzols et sols podzoliques	Spodosols
Sols à sesquioxydes et à matière organique rapidement mineralisée	(several orders)
Sols halomorphes	Aridisols
Sols hydromorphes	(several orders)

TABLE 2.10

Approximate equivalents between FAO soil orders and U.S. Comprehensive System soil orders

FAO	U.S. Comprehensive System
Acrisols	Ultisols
Andosols	Inceptisols
Arenosols	Entisols, Ultisols
Cambisols	Inceptisols
Chernozems	Mollisols
Ferralsols	Oxisols
Fluvisols	Entisols, Inceptisols
Gleysols	Inceptisols, Mollisols
Greyzems	Mollisols
Histosols	Histosols
Kastanozems	Mollisols
Lithosols	(Lithic subgroups of several orders)
Luvisols	Alfisols
Nitosols	Alfisols, Ultisols
Phaeozems	Mollisols
Planosols	(several orders)
Podzols	Spodosols
Podzoluvisols	Alfisols
Rankers	Inceptisols
Regosols	Entisols
Rendzinas	Mollisols
Solonchaks	Aridisols, Mollisols, Inceptisols
Solonetz	Aridisols, Alfisols, Mollisols
Vertisols	Vertisols
Xerosols	Aridisols
Yermosols	Aridisols

REFERENCES

Aubert, G., 1968. Classification des sols utilisée par les pédologues français. In: *Approaches to Soil Classification.* FAO, Rome, World Soil Resources Reports, No 32: 78—94.

Baldwin, M., Kellogg, C.E. and Thorp, J., 1938. Soil classification. In: U.S. Department of Agriculture, *Soils and Man. Yearbook of Agriculture. 1938.* U.S. Government Printing Office, Washington, D.C., pp.979—1001.

Buol, S.W., Hole, F.D. and McCracken, R.J., 1973. *Soil Genesis and Classification.* The Iowa State University Press, Ames, Iowa, 360 pp.

FAO/UNESCO, 1971. *Soil Map of the World, 1:5,000,000, Volume IV, South America.* UNESCO, Paris, 193 pp., 2 maps.

FAO/UNESCO, 1974. *Soil Map of the World, 1:5,000,000, Volume I, Legend.* UNESCO, Paris.

Jenny, H., 1961. *E.W. Hilgard and the Birth of Modern Soil Science.* Collana Della Revista "Agrochimica", Instituto di Chimica Agraria dell' Universita, Pisa, 144 pp.

Kellogg, C.E. and Orvedal, A.C., 1969. Potentially arable soils of the world and critical measure for their use. *Adv. Agron.*, 21: 109—170.

Marbut, C.F., 1927. *The Great Soil Groups of the World and Their Development.* Edwards, Ann Arbor, Mich., 235 pp.

Northcote, K.H., 1971. *A Factual Key for the Recognition of Australian Soils.* Rellim Technical Publications, Glenside, S.A., 123 pp.

Rozov, N.N. and Ivanova, E.N., 1968. Soil classification and nomenclature used in Soviet pedology, agriculture and forestry. In: *Approaches to Soil Classification.* FAO, Rome, World Soil Resources Reports, No. 32: 53—77.

Soil Survey Staff, 1951. *Soil Survey Manual. U.S. Department of Agriculture Handbook No. 18.* U.S. Government Printing Office, Washington, D.C., 503 pp.

Soil Survey Staff, 1960. *Soil Classification — A Comprehensive System — 7th Approximation.* Soil Conservation Service, U.S. Department of Agriculture, Washington, D.C., 265 pp.

Soil Survey Staff, 1976. *Soil Taxonomy.* Soil Conservation Service, U.S. Department of Agriculture, Washington, D.C., in press.

Stace, H.C.T., Hubble, G.D., Brewer, R., Northcote, K.H., Sleeman, J.R., Mulcahy, M.J. and Hallsworth, E.G., 1968. *A Handbook of Australian Soils.* Rellim Technical Publications, Glenside, S.A., 435 pp.

Stephens, C.G., 1962. *A Manual of Australian Soils.* CSIRO, Melbourne, Vic., 3rd ed., 43 pp.

Tyurin, I.V., Gerasimov, L.P., Ivanova, E.N. and Nosin, V.A., (Editors), 1959. *Soil Survey. A Guide to Field Investigations and Mapping of Soils.* (Translated from the Russian by the Israel Program for Scientific Translations, Jerusalem, 1965; also cited as TT65-50062.).

U.S. Department of Agriculture, 1938. *Soils and Man. Yearbook of Agriculture, 1938.* U.S. Government Printing Office, Washington, D.C., 1232 pp.

Chapter 3

CHARACTERISTICS OF ARID-REGION SOILS

INTRODUCTION

Arid-region soils are concentrated between 10° and 50°N and 15° and 50°S (Fig.3.1). Near the equator significant areas are confined to the coastal plains of Peru, northeastern Brazil, and Kenya, Ethiopia, and Somalia in east Africa. The major continuous band of such soils begins on the west coast of Africa and runs across the Sahara and the Arabian peninsula to eastern Mongolia, a distance of more than 13,000 km.

DISTRIBUTION OF SOILS

Of the continents, Africa has the largest area of arid-region soils while Australia has the greatest percentage (Table 3.1). Possibly all of the soils exposed in ice-free Antarctica have characteristics typical of the arid regions but their area is unknown, as is the case in the Arctic.

Dominant soil orders

Entisols are most common, followed by Aridisols, Mollisols, Alfisols, and Vertisols (Table 3.2). The Aridisol order is the only one which is confined, by definition, to the arid regions. All the other orders on the map can be found in both the arid and humid regions although Mollisols are generally restricted to a climate which is semiarid or subhumid and Vertisols must have a long enough dry season for deep cracks to appear in the soil. Alfisols grade into Ultisols in northern and northeastern South America and in Africa. If climate were the sole distinction, some Ultisols would be included among arid-region soils, but they are defined in the U.S. Comprehensive System as belonging to humid regions, and they are so treated here. In the extremely arid climates, only Aridisols and Entisols are of significant extent.

Many people associate the arid regions with shifting, barren sand dunes (Psamments). While sand dunes are common, they are by no means dominant except in local areas, especially in the northern half of the Sahara. The most representative soil, worldwide, is a coarse to medium textured, shallow to moderately deep calcareous Entisol (Orthent suborder) or Aridisol (Orthid suborder) with a desert pavement. A moderate or high base saturation is the one property which arid-region soils have in common.

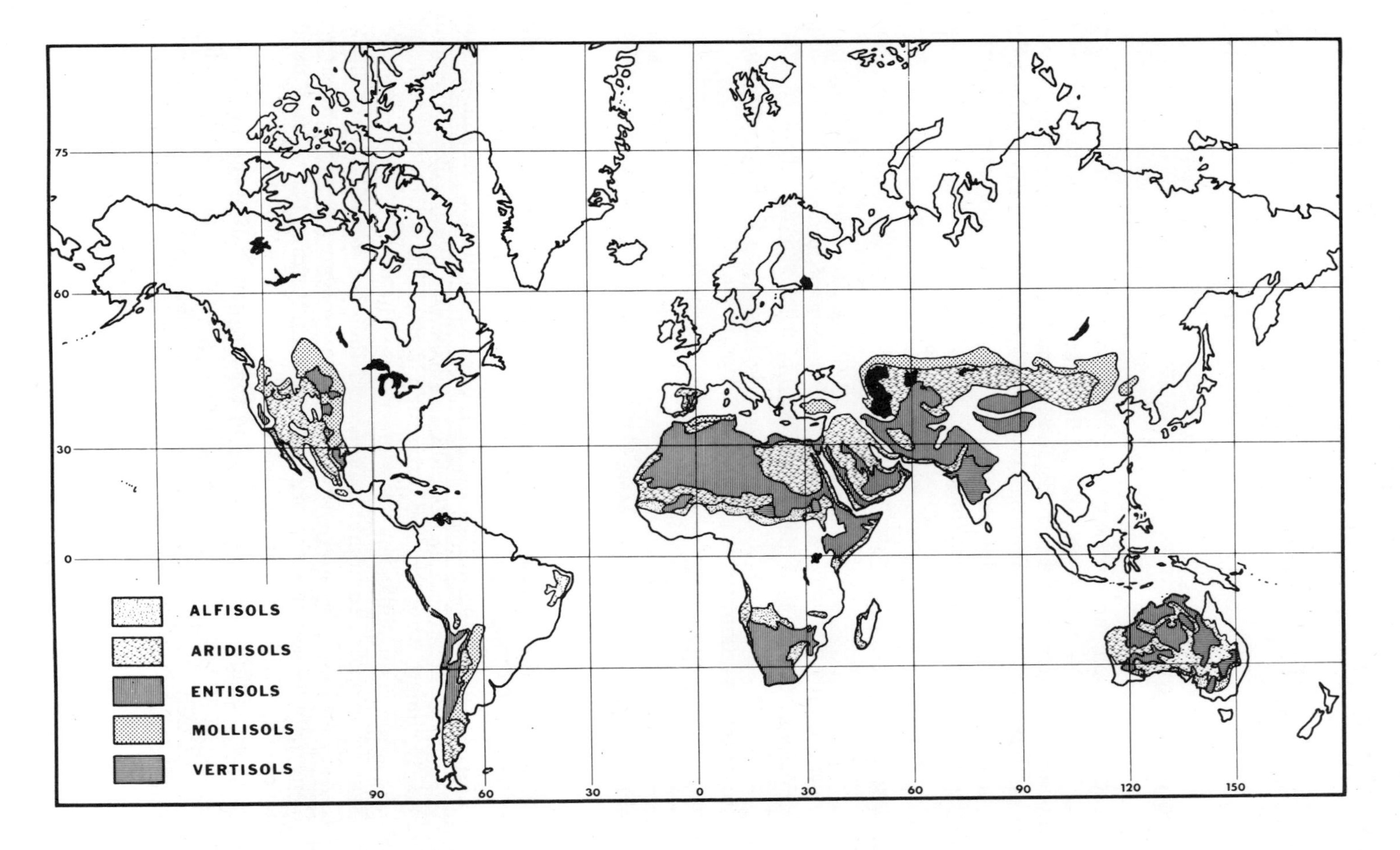

Fig.3.1. Arid-region soils of the world. Polar soils not shown.

TABLE 3.1

Distribution of arid-region soils by continents*

Continent	Arid-region soils	
	area (km^2)	percent of continent
Africa	17,660,000	59.2
Asia	14,405,000	33.0
Australia	6,250,000	82.1
Europe	644,000	6.6
North America	4,355,000	18.0
South America	2,835,000	16.2
Total	46,149,000	

*Excluding polar regions.

TABLE 3.2

Approximate area of arid-region soils of the world, by soil order*

Soil order	Area (km^2)	Percent of arid-region soils	Percent of world land area
Alfisols	3,070,000	6.6	2.1
Aridisols	16,570,000	35.9	11.3
Entisols	19,149,000	41.5	13.1
Mollisols	5,475,000	11.9	3.7
Vertisols	1,885,000	4.1	1.3
Total	46,149,000	100.0	31.5

*Excluding polar regions.

Economically, the alluvial soils (Fluvents) along rivers and streams are the most important ones. Virtually all of the 35,000,000 people of Egypt live on a strip of Nile alluvium only a few kilometers wide. That strip of land is so small that it must be exaggerated in size in order to be shown on the map in Fig.3.1. Many other strips of alluvial soils could not even be shown, despite the significant role they play in the local economy, because they are too narrow to be delineated on a small-scale map without excessive exaggeration in size.

Alfisols are in a rather unique category among the soil orders. They occur in two quite different environments: tropical semiarid savannah and temperate humid forests. The arid variant of Alfisols is found where the vegetation is grassland with scattered xerophytic trees, the growing season is long, protracted dry seasons occur some time during the growing season, and the rainy period or periods are short. The Sudanian Zone of Africa south of the Sahara typifies the savannah Alfisol area. Most of the Alfisols in the mountains of North America (Fig.3.1) actually belong to the humid variant of the order.

Comparison of soil and climatic maps

A comparison of the soil map (Fig.3.1) with the climatic map of the arid regions (see Fig.1.1, p.2) discloses differences in boundaries. In drawing the soil map, those soils which extended into the humid regions from the arid regions were continued until a soil boundary occurred. The most obvious example of this is in India, where Vertisols are extensive on the Deccan Plateau. The climate is arid in the west and south and humid in the northeast, but Vertisols are dominant throughout. A similar situation — for Entisols — is found in northern Australia where the climate becomes humid. In that case, the Entisol boundary is the sea coast. On the other hand, the Tibetan Plateau in China is not shown in Fig.1.1 as having an arid climate, yet it is delineated in Fig.3.1 as having arid-region Entisols. The reason for including the Tibetan Plateau in Fig.3.1 is that the soil map of Asia prepared by Kovda and Lobova (1971) has mountain-desert soils and saline flats appearing there. It seems quite possible, in view of the high altitude of the plateau and the low temperatures, that soils approximating cold-desert soils can be found in the plains between the mountain ranges. Whether or not this is true remains to be seen.

DESCRIPTIONS OF ORDERS AND SUBORDERS

Arid-region soils fall, mainly, within five orders and thirteen suborders (Table 3.3). Ustalfs and Xeralfs are the major Alfisol suborders; both Argids and Orthids are common among the Aridisols; Orthents and Psamments dominate the Entisol order; Ustolls stand out among the Mollisols; and Ustert is the usual suborder of the Vertisols.

Brief descriptions of the orders and suborders of Table 3.3 are given here to providc a general indication of the principal distinguishing properties of each one. Complete definitions can be found in *Soil Taxonomy* (Soil Survey Staff, 1976).

Alfisols

Alfisols representative of the central concept of the order have an ochric epipedon, an argillic horizon, moderate to high base saturation, and enough moisture to keep the moisture level above the wilting point for at least three months each year during the growing season. Other Alfisols may have a

TABLE 3.3

Orders and principal suborders of arid-region soils

Order	Suborder
Alfisol	Aqualf
	Ustalf
	Xeralf
Aridisol	Argid
	Orthid
Entisol	Aquents
	Fluvents
	Orthents
	Psamments
Mollisol	Boroll
	Ustoll
	Xeroll
Vertisol	Ustert

fragipan, duripan, natric horizon, petrocalcic horizon, plinthite, or other horizons. All Alfisols have an argillic or natric horizon. The moisture regime normally is aquic, udic, ustic, or xeric but may be aridic if the epipedon is both massive and hard or very hard when the soil is dry. Base saturation is 35% or more in the lower part of the profile. A mollic epipedon is permitted in Alfisols if the base saturation in the argillic horizon is less than 50%. There may not be a spodic or oxic horizon overlying an argillic horizon. In regions where the soil temperature regime is thermic or warmer, Alfisols tend to be found in the savannahs between Aridisols and the Ultisols and Oxisols of the warm humid regions. Where cooler temperature regimes prevail, Alfisols commonly form a belt between Mollisols and the Spodosols and Inceptisols of the cold humid regions.

Aqualfs are Alfisols showing evidence of poor drainage (gray color and mottling in the subsoils) and having an aquic soil moisture regime unless drained artificially. A fluctuating water table seems to be essential to the development of Alfisols, with the water table sometimes dropping below the argillic horizon and at other times being perched above it.

Ustalfs are the dry Alfisols. The soil moisture regime is ustic and the soil temperature regime may be mesic, isomesic, or warmer but usually is at least as hot as thermic. Soil color usually is reddish; drainage is good. A calcic horizon sometimes is found in or below the argillic horizon. These are the soils of the savannahs where xerophytic trees and grass provide the vegetative cover and where the rains are concentrated during the summer period.

Xeralfs are the Alfisols of Mediterranean climates. The moisture regime is xeric. While the soils are dry for extended periods during the summer, there is considerable moisture in the soil during the wet winters. Temperature regimes are warmer than cryic and cooler than hyperthermic, usually being thermic.

Aridisols

Aridisols are mineral soils of the arid regions. They have a low organic-matter content. During most of the time when temperatures are favorable for plant growth, the soils are dry or salty, with consequent restrictions on growth. During the warm season, there is no period of three months or more when soil moisture is continually available to plants, except in places where a water table is close to the surface. Aridisols must have one or more pedogenic horizons which may have formed under present climatic conditions or under earlier pluvial conditions.

There is no oxic or spodic horizon in Aridisols. There is an ochric (light-colored) or anthropic (mollic-like) epipedon and the soils have one or more of the following: (1) an argillic or natric horizon, an epipedon which is not both hard and massive when dry, and an aridic moisture regime; (2) a salic horizon with its upper boundary within 1 m of the surface; or (3) no argillic or natric horizons but a calcic, petrocalcic, gypsic, petrogypsic, or cambic horizon or a duripan (singly or in combination) within 1 m of the surface, along with an aridic moisture regime or an ustic or xeric moisture regime if the salinity of the saturation extract is 2 mmhos or more in the upper part of the profile.

The temperature regimes of Aridisols vary from cryic to isohyperthermic (cold deserts to hot deserts). Vegetation consists of shrubs, forbs, and grasses, sparsely distributed on the uplands but sometimes fairly dense in depressions.

A common characteristic of upland Aridisols is a desert pavement of pebbles, gravels or stones. The pavement usually is the result of the removal of fine material by water or wind, leaving the larger-sized fraction on top of the soil. Among other things, the pavement helps reduce subsequent erosion by protecting the underlying soil.

Argids have an argillic or a natric horizon in which silicate clays have accumulated by illuviation from upper horizons. They may also have a calcic or petrocalcic horizon or a duripan. Argids may occur on gentle or steep slopes and may be calcareous or non-calcareous. Their presence in an arid region indicates, in general, that they are on old surfaces (Late Pleistocene or older) since clay translocation would be slow under an arid climatic regime. Much of the clay illuviation may have occurred in pluvial periods antedating the present arid climate.

Orthids do not have an argillic or natric horizon. They do have one or more of the following pedogenic horizons or layers: calcic, petrocalcic, gypsic, petrogypsic, salic, or cambic horizon or a duripan layer. Orthids are developed on younger materials than the Argids, with the single exception of those having a petrocalcic horizon which has developed in the argillic horizon of a previous soil. It is likely, then, that all but the petrocalcic Orthids are the product of the present climate.

Entisols

Entisols are mineral soils showing little or no evidence of development of pedogenic horizons. Many of the Entisols have an ochric epipedon and some have an anthropic epipedon. A few of the sandy ones have an albic horizon and those in coastal marshes have a histic epipedon. Entisols do not have well-defined pedogenic horizons but many of the alluvial Entisols (Fluvents) are stratified with layers of differing textures. Pedogenic horizons have not formed because, primarily, the soils are too young due to recent deposition of fresh material or to eroding away of the previous surface. Entisols may have any temperature or moisture regime except the combination of very low temperatures and wetness. Entisols may be underlain at depths greater than 50 cm by buried soils (paleosols).

Clays displaying wide cracks (> 1 cm) at a depth of 50 cm or showing some indication of pedogenesis are classed as Vertisols rather than Entisols, even though there are no pedogenic horizons present. Vertisols are common in depressions (playas) in arid regions.

Aquents are wet Entisols of tidal marshes, deltas, flood plains of streams, or in wet very sandy deposits. Soil colors are bluish or gray, with mottles. They may have any temperature regime except pergelic (continuously cold) and the moisture regime is aquic or peraquic (continuously wet).

Fluvents are found in recent water-deposited sediments, principally in flood plains, fans, and deltas of rivers and streams where drainage is not poor. Stratification of soil materials is normal. Flooding occurs frequently. Fluvents have a texture of loamy fine sand or finer below the Ap (cultivated) horizon, if one is present, or below 25 cm but above 1 m. Organic-matter content decreases irregularly with depth, due to stratification of the soil. The epipedon is ochric or anthropic. The slope of the land must be less than 25% and is usually level.

Orthents are undeveloped soils on (1) surfaces where erosion has removed completely any horizons which would be diagnostic for other orders than Entisols, or (2) saline substrates, usually gypsiferous, where a water table is deep or absent. Orthents can be found in any climatic region. They do not occur where there is a high water table or on shifting or stabilized sand dunes. Their most common occurrence is on rocky slopes.

Psamments are sandy Entisols. They may occur under any climate and on surfaces of any age. They have a sandy texture in all subhorizons below the Ap horizon, if present, or below 25 cm and above 1 m. Psamments contain less than 35% gravel or coarser fragments, in all subhorizons, and they are not permanently saturated with water and do not have properties of Aquents. Gravelly sandy soils are Orthents rather than Psamments. Included in the Psamments are unstabilized sand dunes (shifting dunes), stabilized sand dunes, cover sands, and sand fields. Vegetative cover ranges from none on the shifting dunes to good grass growth on stabilized sand fields. Wind erosion is a major hazard.

Mollisols

Mollisols are found mainly in the semiarid and subhumid grasslands of the world where the soils have dark-colored surfaces and are rich in bases such as calcium and magnesium. Practically all of them have a mollic epipedon and many have an argillic, natric, or calcic horizon. A few Mollisols have an albic horizon and some have a duripan or a petrocalcic horizon. Mollisols may have any of the temperature or moisture regimes but growing conditions must be adequate to support perennial grasses. This order consists of mineral soils having either a mollic epipedon or a surface horizon that meets all the requirements for a mollic epipedon, except for thickness, after mixing the upper 18 cm. If there is an argillic horizon, it must have a base saturation of 50% or more in and below the argillic horizon. If there is a cambic horizon, base saturation from the surface to a depth of no more than 1.8 m must be 50% or greater. The upper part of the soil (above 35 cm) cannot be dominated by volcanic ash. There are additional restrictions on calcium carbonate content, montmorillonitic-clay content, and depth of cracks if the soil temperature regime is warm or hot. Mollisols do not have an oxic horizon nor do they have a spodic horizon within 2 m of the surface.

Borolls are more or less freely drained soils of cool to cold continental climates. They have a mean annual soil temperature lower than 8°C and do not have a xeric moisture regime unless the temperature regime is cryic. Any albic, argillic, or natric horizon does not have mottles or iron—manganese concretions more than 2 mm in diameter. Borolls usually have an ustic moisture regime.

Ustolls are Mollisols of the summer rainfall semiarid and subhumid climates. Their temperature regime is warmer than frigid. They are more or less freely drained, do not show evidence in their profiles of poor drainage, and are not saturated with water at any period in the year. Their moisture regime is ustic or aridic but cannot be xeric. If the moisture regime is wetter than ustic, a soil meeting the other criteria will qualify as an Ustoll when there is a calcic or gypsic horizon or a calcareous layer below any cambic or argillic horizon which may be present. Presence of lime or gypsum accumulations is enough to indicate that the soil is dry for considerable lengths of time.

Xerolls are Mollisols formed under a Mediterranean climate of wet winters and dry summers, the opposite of the climate under which Ustolls form. The mean annual soil temperature is lower than 22°C and mean summer and mean winter soil temperatures differ by at least 5°C at a depth of 50 cm. Internal drainage is sufficient to avoid development of mottles or iron—manganese concretions larger than 2 mm in diameter. Xerolls have a xeric moisture regime or an aridic regime which borders on being xeric. They do not have a cryic temperature regime.

Vertisols

Vertisols are the moderately deep and deep cracking clays of the warm regions. Clay content is 30% or more in all subsections of the profile down to 50 cm. They have a mesic, isomesic, or warmer soil temperature regime and must be dry enough at some time in most years for cracks at least 1 cm wide to extend to a depth of 50 cm below the surface. There is no bedrock, petrocalcic horizon, or duripan within 50 cm of the surface. The soils frequently are slightly to moderately calcareous throughout the profile. Many Vertisols occupy a type of microtopography called "gilgai", where small mounds, usually less than 50 cm high, are scattered around the level landscape in a more or less regular fashion (Hallsworth and Beckmann, 1969). The mounds form as the result of surface soil falling into cracks and exerting pressure on the soil column when the cracks close after the soil is wetted again. The pressure is relieved by upward expansion of the soil in the vicinity of the cracks. Vertisols are generally found on level surfaces, but may occur in undulating or moderately sloping topography. They may form in alluvium or in place from parent rock.

Usterts are distinguished by cracks that remain open for 90 consecutive days or more in most years but are closed for at least 60 consecutive days when the soil temperature at 50 cm is above 8°C. The cracks usually open and close more than once in most years. The mean annual soil temperature is 22°C or more or the mean summer and mean winter temperatures at 50 cm differ by less than 5°C (an iso-temperature pattern). Usterts are found where there are (1) monsoon climates; (2) two rainy and two dry seasons in tropical or subtropical zones; or (3) temperate zones with low summer rainfall.

DISTINGUISHING FEATURES OF ARID-REGION SOILS

Several soil-geomorphology associations serve to distinguish the arid regions from the humid regions, as do certain chemical properties of the soils. Aridity is, of course, the controlling factor but age of landscape, type of geologic material, distribution and intensity of precipitation, and man's alteration of the physical and biological environment modify — for better or for worse — the effect of aridity.

Sand dunes

Sand dunes are not restricted to the arid regions but they are a characteristic component of the landscape in those regions. Deserts usually are envisioned, in the popular mind, as a vast sea of sand stretching endlessly to the horizon — barren, hot, and waterless. Some dune fields do fit that description and are called, aptly enough, sand seas. Among the best examples are the ergs of North Africa, especially the Grand Erg Oriental, Grand Erg Occidental, and Erg Chech of Algeria, the Rub' al Khali of Saudi Arabia, and the Takla Makan Desert of China. In these dune fields, a continuous or nearly continuous layer of sand covers the underlying rock or sediments. Stabilizing barren dunes in arid regions is very difficult, whether some mechanical or chemical treatment is given them or whether vegetation establishment is attempted. Lack of water is part of the problem in vegetation establishment; preventing the covering of plants or their destruction from the blasting effect of sand during wind storms also is essential.

Not all large sand dunes are barren of vegetation. In the better-watered places, trees, shrubs, and grasses help anchor the sand. Moisture sometimes is more available than one might think because any rain which does penetrate the sand to a depth of 30 cm or more tends to be protected from evaporation as the sand above it dries out and forms a barrier to further evaporation.

Sand dunes come in all sizes and shapes. They vary in height from 30 cm to over 100 m and in distribution from an isolated dune on a clay plain to a continuous series covering an area of thousands of square kilometers. There are longitudinal dunes lined up parallel to the dominant wind direction, transverse dunes in rows perpendicular to the wind, barchan dunes having the familiar crescent shape with the horns pointed downwind, dome dunes having a roughly circular shape, and others. Coppice dunes are those formed around shrubs, and generally are no more than a meter or two in height.

Moving or shifting sand dunes are fairly uniform in texture with depth, due to the continual mixing as the dunes move across the landscape. Stabilized dunes may show the beginning of horizonation as fine material and carbonates in the surface soil are carried downward in percolating water. Soil pH usually approaches neutrality or may be slightly acid in the surface, then becomes alkaline in the lower depth.

Dune material varies in composition although all dunes usually are called sand dunes. The White Sands of New Mexico are nearly pure gypsum and the kopi dunes on the leeward sides of saline depressions in Australia are largely gypsum. Quartz and other minerals are, however, the commonest components of sand-sized particles.

Desert pavement

Desert pavement is another characteristic feature of arid-region soils. The pavement consists of an unconsolidated layer of gravel, pebbles, or stones lying on top of the soil. It is also referred to as desert armor because it protects the underlying soil from wind and water erosion. The names used in Africa for a surface covered with desert pavement are reg and hamada; in Mongolia and China the term is gobi.

The formation of desert pavement is due to one or more of the following processes: (1) removal of fine material by wind erosion; (2) removal of fine material by water erosion; (3) upward movement of gravel and stones due to alternate welting and drying (expansion and contraction of the fine material between the coarse particles); (4) deposition of gravel and stones by water; and (5) weathering of surface crusts of silica or calcium carbonate (Fig.3.2). The upper (exposed) side of the particles sometimes is darkened and polished with what is called desert varnish, apparently consisting of iron and manganese oxides dissolved out of the pebbles and precipitated on the pebble surface. Undersides of pebbles frequently have white or gray calcium carbonate coatings where they are in contact with the soil. Desert pavements are most likely to be found on gravelly and stony soils.

The presence of surface pebbles serves not only to protect the soil from further erosion but also reduce moisture losses. The pebbles slow down water movement across the surface, which results in more water penetrating the soil, and also reduce the amount of soil surface exposed to evaporation.

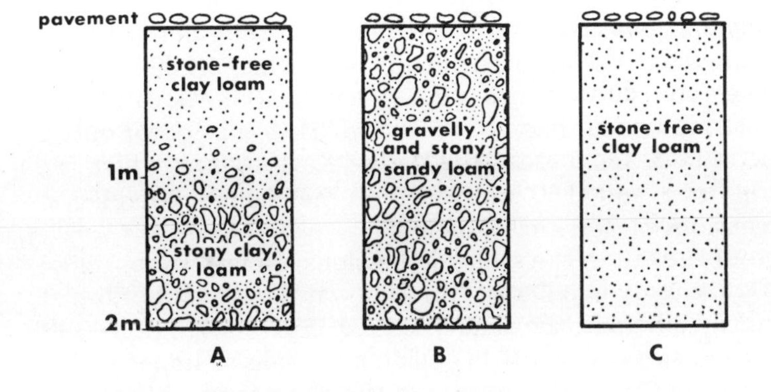

Fig.3.2. Soil profiles underlying desert pavement. Pavement may form by (*A*) stones moved upward by soil expansion and contraction; (*B*) stones remaining on surface after removal of finer soil particles by wind and water; or (*C*) stones formed on surface from weathering of exposed indurated soil layer or continually moved upward through periodic deposits of fine-textured material.

Playas

A third characteristic feature of arid-region soils is the presence of playas. These are broad, usually shallow depressions in the landscape, from which there is no surface outflow of water unless exceptional storms occur. The depressions constitute closed basins in which clay and silt particles — and some sand — accumulate as the result of water erosion on the surrounding uplands. Playas are found mostly in mountain-and-basin landscapes, but many playas are only shallow depressions in broad plains. Playas commonly have saline or saline—sodic soils; playas are called *takyrs* in the Soviet Union.

Takyrs are medium to fine textured, slowly permeable soils, without a water table close enough to the surface to permit salt crusts to appear, have a generally nonsaline surface to a depth of a few centimeters, a slightly to moderately saline subsoil, and — most importantly — a high content of exchangeable sodium in the surface (Lobova, 1960). The surface usually is completely devoid of higher plants but frequently is covered with algae and lichens when wet. Water stands on the surface for long periods following rains and provides a source of water for humans and animals during dry seasons. Soil permeability is sufficient, however, to permit salts to be leached out of the surface.

A distinguishing characteristic of takyr soils is the polygonal-shaped cracks which appear on the surface when the soil dries. Polygons are typical features of fine-textured, sodium-affected, nonsaline soils, with the polygons having a diameter of about 20—30 cm and a convex surface between the cracks.

Saline depressions

Similar to the takyrs in physiography but different in the salinity of the soil are the saline depressions which occur nearly everywhere in the hot and cold arid regions. They are called salinas, salares, sebkhas, chotts, salt lakes, salt pans, and salt flats, in various parts of the world. They may cover only a few hectares of land or be as immense as the Dasht-i-Kavir in Iran, which is a basin about 500 km long and 150 km wide, or the Salinas Grandes in Argentina, with dimensions of 230 km by 50 km.

The major source of the salt in the saline depressions is a water table close to the surface. Water rising by capillarity from the water table evaporates at the soil surface and leaves behind the salts dissolved in the water. Some salt enters the depressions in surface runoff but seldom are soils so impermeable that a salt crust would appear on the surface in the absence of a high water table.

Gypsum is a common constituent of the salt crusts of saline depressions. Being soluble to the extent of only about 2,000 ppm, gypsum is less likely to be leached out of the upper part of the soil than is, for example, a highly soluble salt such as sodium chloride. The gypsum "sand" dunes of the world appear on the leeward side of salt flats, from which gypsum crystals are blown out during dry periods.

Tablelands

Tablelands (mesas) form another special feature of arid-region landscapes. As used here, the term refers to an isolated, flat-topped, table-like mountain formed from a plateau. The table-like appearances persist because erosion is minimal on the isolated mountains, capped as they are with erosion-resistant rocks, usually sandstone or basalt. It is on these tablelands, isolated for tens or hundreds of thousands of years, where argillic Aridisols appear. The old landscapes are ideal for demonstrating how the very slow process of clay illuviation in an arid region can, in time, produce a distinct argillic horizon, if there is no interference from erosion or deposition. It is on the tablelands, too, where rock disintegration leads to a layer of angular stones and rocks accumulating on the surface (a type of desert pavement), as in Australia where the indurated siliceous B horizon of a former soil is brought to the surface by erosion of the A horizon.

Wadis

Wadis, also called oueds, nullahs, or arroyos, are the drainage channels for intermittent streams. Customarily, they have vertical side walls and gravelly or sandy beds. It is not unusual to have a dry wadi turn suddenly into a raging torrent following a rain in the upper part of the watershed. The sudden influx of water to the wadi frequently causes the flow to appear as a wall of water rather than as a gradually increasing flow. Unwary travellers have learned to their sorrow that the bed of a wadi is no place to be caught when the crest of a flash flood arrives. The erosive power of a high-intensity, short-duration desert storm is great; it accounts for the often disastrous destruction of property and deposition of sand, gravel, and boulders on lower-lying areas. Normally, wadi beds absorb much of the runoff from surrounding lands, especially when rains are only moderately intensive, and little of the water may reach the mouth of the wadi.

Indurated horizons

Indurated calcium carbonate and gypsum represent physical features unique to arid regions. Indurated carbonate layers are called petrocalcic horizons or calcrete and indurated gypsum layers are petrogypsic horizons. Both types of horizons are subsurface horizons, normally, but may be exposed on the surface due to erosion processes. They are pedogenic horizons, not geologic material, and can usually be recognized as pedogenic by their closeness to the soil surface and their relative thinness (1 or 2 m thick, at most) over different materials. Carbonates and gypsum accumulate because they are the two least soluble of the salts commonly found in soils. They become indurated when freshly precipitated carbonate or gypsum occupies the spaces between nodules formed previously and, on dehydration, cements the nodules together. Petrocalcic horizons are widespread, world-wide, on relatively old surfaces; petrogypsic horizons are found in Algeria, Tunisia, and the U.S.S.R., and probably in other countries as well.

Silcrete (indurated siliceous) horizons are a prominent part of arid-region soils on old landscapes in Australia. They are believed by some to be relics of a previous, more pluvial, climate rather than being the product of soil-forming processes in an arid climate (Jackson, 1957). Silcrete is almost entirely secondary silica. It does not appear to be common in soils outside Australia although petrocalcic horizons containing significant amounts of silica have been found in the United States, South Africa, and elsewhere.

Base saturation

The one chemical property arid-region soils have in common is a moderate to high base saturation. Under low-rainfall conditions, leaching of basic cations (calcium, magnesium, potassium, and sodium) is slow and there is no opportunity for acidic cations (hydrogen and aluminum) to accumulate on the cation-exchange complex. A high base saturation means a soil pH above 7; a moderate base saturation generally results in a soil pH of 5.5 or more.

REFERENCES

Hallsworth, E.G. and Beckmann, G.G., 1969. Gilgai in the Quaternary. *Soil Sci.*, 107: 409—420.

Jackson, E.A., 1957. Soil features in arid regions with particular reference to Australia. *J. Aust. Inst. Agric. Sci.*, 25: 196—208.

Kovda, V.A. and Lobova, E.V. (Editors), 1971. *Soil Map of Asia.* USSR Academy of Sciences, V.I. Lenin All-Union Academy of Agricultural Sciences, Moscow, scale 1: 6,000,000.

Lobova, E.V., 1960. *Pochvy Pustynnoi Zony SSSR.* (*Soils of the Desert Zone of the U.S.S.R.* Issued in translation by the Israel Program for Scientific Translations, Jerusalem, 1967, 405 pp; also cited as TT67-51279.)

Soil Survey Staff, 1976. *Soil Taxonomy.* Soil Conservation Service, U.S. Department of Agriculture, Washington, D.C., in press.

Chapter 4

AFRICA

INTRODUCTION

Africa has two well-defined arid zones (Fig.4.1). The Sahara, in the minds of many the epitome of the world's deserts, dominates the northern zone stretching 7,500 km across the continent from the Atlantic Ocean to the Arabian Sea and extending 2,800 km from north to south in the middle. The Sahara itself is about 5,500 km from east to west and 1,800 km from north to south through Algeria and Niger. The southern arid zone occupies most of the southern tip of the continent and includes the west coast of the island of Madagascar.

Entisols and Aridisols are by far the most extensive soil orders, followed by a lesser area of Alfisols and a small area of Mollisols on the north coast (Fig.4.2). Most of the information on soils was taken from the publication by D'Hoore (1964) and supplemented by maps for South Africa and some of the former French territories.

ENVIRONMENT

Climate

Two radically different rainfall distribution patterns are found in the African arid zones: (1) wet winters and dry summers along the Mediterranean coast and in South Africa and (2) dry winters and a short wet summer season south of the Sahara. There are variations from this pattern, including that of practically no rain at any time in the central Sahara, but one or the other climatic type dominates much of the continent's dry regions. Rainfall effectiveness, insofar as it affects soil properties and agricultural production, differs greatly between the two rainfall patterns. Rain during cool winters leads to maximum effectiveness of precipitation since evaporation rates are relatively low and water penetrates to greater depths in the soil. Consequently, soils of Mediterranean climates are leached more and they tend to be non-saline and less calcareous in the upper horizons than their counterparts in the summer-rainfall area. Furthermore, rains in Mediterranean climates are gentler than summer thunderstorms, and less surface runoff and soil erosion occur. A mean annual rainfall of 400 mm in North Africa, spread out over several winter months, is adequate to produce a good wheat crop whereas 400 mm

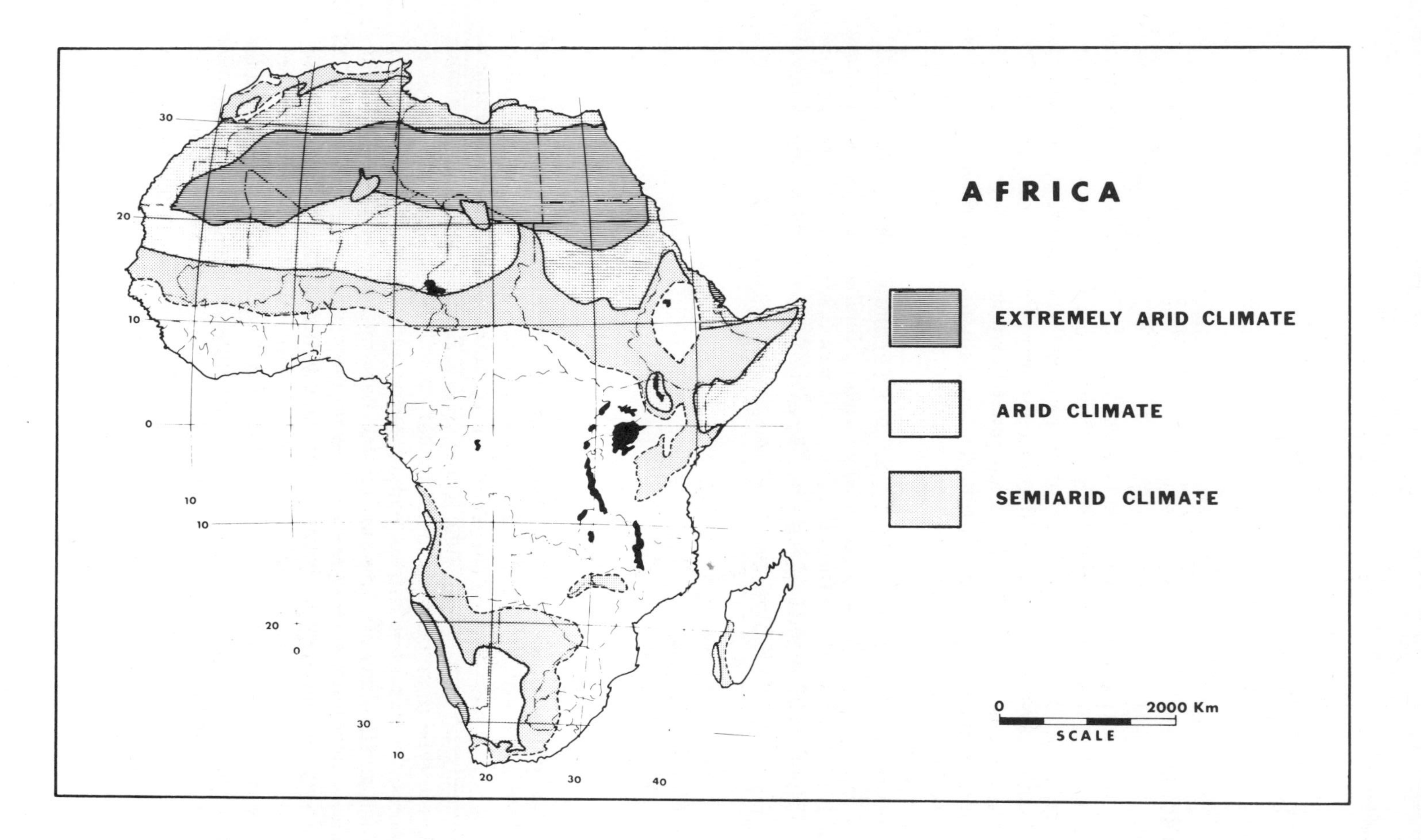

Fig.4.1. Arid regions of Africa (after Meigs, 1953).

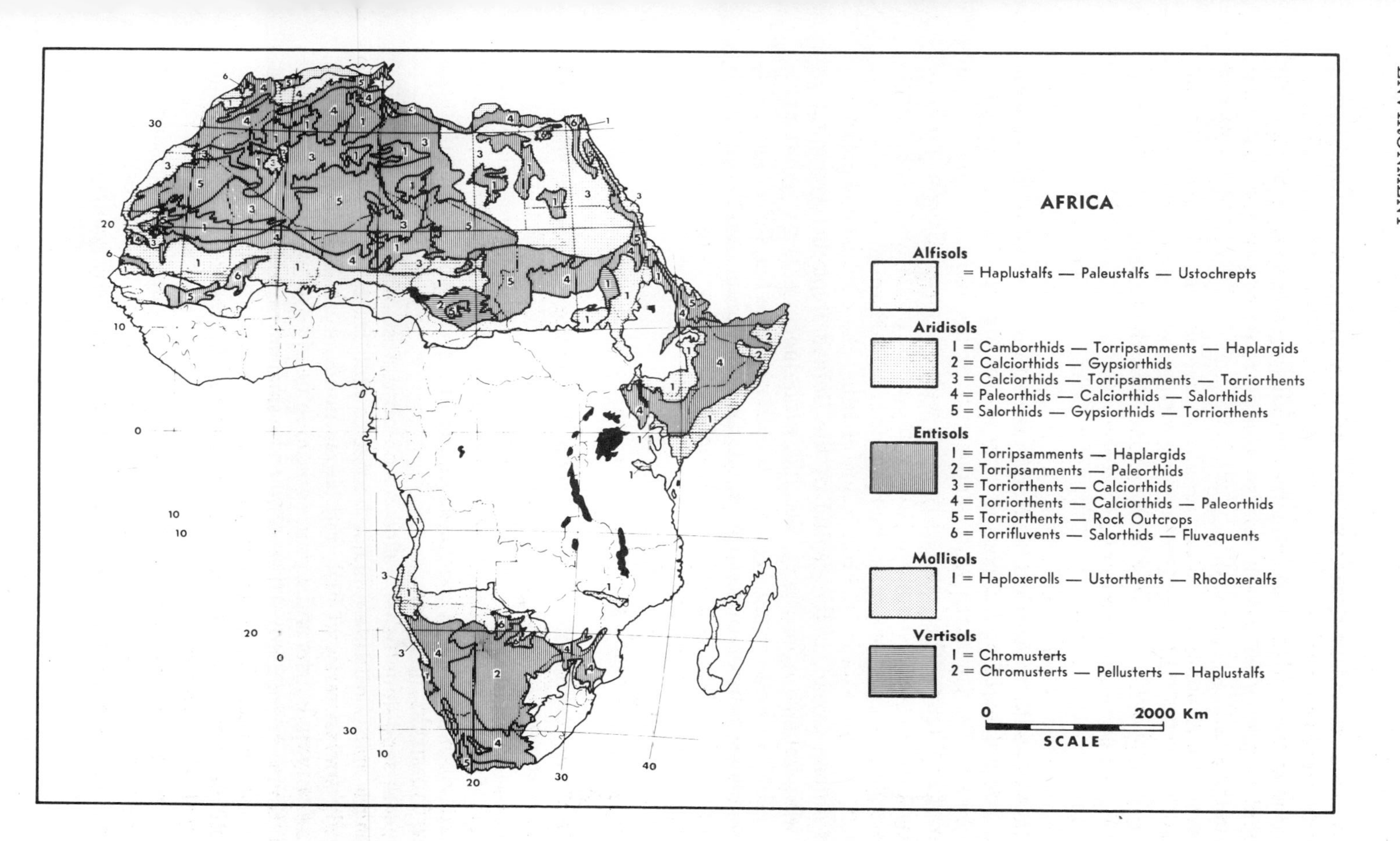

Fig.4.2. Arid-region soils of Africa.

falling in July and August in Mali may be only marginally adequate for a moderate sorghum crop.

Starting in the center of the Sahara, rainfall increases to the north and to the south (Fig.4.1). East—west rainfall zones are narrow in the north but broad in the south, where the driest zone is the Saharan, followed in order of increasing annual precipitation and length of the summer wet season by the Sahelian Zone, the Sudanian Zone, and the Guinean Zone on the coast. On the map in Fig.4.2, arid-region soils extend into the Sudanian Zone where Alfisols and a savannah type of plant cover are found.

In southern Africa, the tip of the continent has wet winters and dry summers but the pattern changes rapidly to summer rains in the north. The driest part of southern Africa is the Namib Desert on the coast of South West Africa. The Kalahari Desert is considerably wetter than the Namib. Precipitation increases to the north, east, and south of the Kalahari, going from 150 mm to 500 mm. The Kalahari is much wetter than the Sahara and has a lush vegetative cover in comparison to the amount of vegetation in the Sahara.

Climatological data for five weather stations in Africa are given in Table 4.1. The first four are in a north—south line crossing the continent; Mogadishu, on the Indian Ocean, is the capital of the Somali Republic. Precipitation varies with distance from coasts, altitude, wind direction, and ocean temperatures. Adrar, in west central Algeria, is the driest of the five stations and also has the greatest temperature range: an all-time maximum of 51°C and minimum of —4°C. Of the other stations, only Kimberley ever experienced a minimum temperature below 0°C.

Geology and geomorphology

Precambrian crystalline rocks underlie the entire continent. They are exposed over about one-quarter to one-third of the land surface in the African arid regions. Most of the land is covered with sandstone, limestone, loose sands, alluvium and colluvium. Mountain chains are located in the northwest, east, and south sections of the continent, and there are isolated massifs (e.g., Hoggar, Tibesti) in the Sahara.

Immense sand-dune fields, called *ergs* in northwest Africa, are a feature of the Sahara and constitute a major barrier to land travel because they are a seemingly endless series of high and barren dunes. Apart from the ergs, sand dunes of varying height and area are common nearly everywhere in northern Africa. Dunes in southern Africa tend to be smaller, except for the southern half of the coastal Namib Desert in South West Africa, than in the north, probably because the greater vegetative cover in the south reduces the amount of wind erosion.

A second major feature of northern Africa arid regions is the presence of extensive surfaces with a well-defined desert pavement. The soil beneath the pavement of gravel may be coarse or medium textured and generally is fairly

TABLE 4.1

Climatological data for five African weather stations

Station	Elevation (m)	Precipitation (mm)													Average annual temperature (°C)
		J	F	M	A	M	J	J	A	S	O	N	D	annual	
Algiers (Algeria)	59	112	84	74	41	46	15	0	5	41	79	130	137	764	18
Adrar (Algeria)	286	0	0	2	0	0	0	0	0	0	5	5	0	12	24
Niamey (Niger)	216	0	0	5	8	33	81	130	188	94	13	0	0	552	29
Kimberley (South Africa)	1,119	61	64	79	38	18	5	5	8	15	25	41	51	410	18
Mogadishu (Somali Rep.)	12	0	0	0	58	58	96	64	48	25	23	41	13	426	27

Source: U.S. Department of Commerce (1969).

deep over bedrock. When the desert pavement mantles a smooth level or gently undulating surface such as alluvial fans and piedmont plains, the surface is called a *reg* by French scientists. The desert pavement of regs is formed by the removal of finer soil material by wind and water erosion or by the deposition of gravels on the soil surface as the result of erosion from higher slopes. The former are called residual regs and the latter transported regs. Vehicular travel across regs is easy because the desert pavement is dense enough to provide a relatively durable surface.

Dry salt lakes (sebkhas, chotts) are most numerous in a broad zone extending from Tunisia to Mauretania. The largest single salt sink on the continent is the Qattara Depression in northwest Egypt.

Vegetation

Three vegetation zones are dominant in the African arid regions according to D'Hoore (1964): (1) a desert zone devoid of plants except for a few scattered trees or shrubs; (2) a steppe zone with a sparse vegetative cover of grass, shrubs, and dwarf trees; and (3) a savannah of grassland and scattered trees and shrubs. The desert zone does not occur in southern Africa to any significant degree. Alfisols are the soils of the savannahs; Aridisols, Entisols, and Vertisols are the usual soils in the other vegetative zones.

Acacia species are by all odds the commonest and most widespread trees of arid Africa. They are found on all kinds of soils and topographic positions. In addition to *Acacia*, there are trees and shrubs of *Adansonia*, *Balanites*, *Capparis*, *Commiphora* and *Euphorbia*, with *Tamarix* along waterways. Common grasses include *Aristida*, *Schoenefeldia*, *Cenchrus*, *Cymbopogon*, and *Eragrostis*, with *Andropogon*, *Sporobolus*, and *Juncus* on the inundated clay plains (Vertisols). Two introduced species have become of increasing importance, especially on the edge of the Sahara. One is the *Eucalyptus* from Australia, which does well in the revegetating of sand dunes; the other is mesquite (*Prosopis juliflora*) from the western hemisphere, a leguminous shrub which grows well in arid regions and produces edible and nutritious seeds in its large pods.

Destruction of vegetation in South Africa during the 19th and 20th centuries led to alarm about the adverse effects it was having on people, livestock, and soil and water resources (Dyer, 1955). In 1914 the South African parliament established a Select Committee to study the problem. It reported on rainfall, soil erosion, droughts, and remedial measures. In 1950, after several other studies, a Desert Encroachment Commission was organized following a period of accelerated soil erosion and livestock losses to droughts. The commission concluded that there was no evidence of a significant decrease in rainfall amount or character during historical times and that the responsible agent for the desertification which has occurred was man, in the form of overgrazing and land and water abuse. As Dyer (1955) noted, in view of the changes in soil—plant—water relations following desertification, it is "obviously impracticable to restore completely the flora of 300 years ago" in the affected area. For all practical purposes, the changes are permanent.

SOILS

Entisols, consisting largely of sand dunes and shallow to deep gravelly soils, are by far the most extensive soils in the arid regions of Africa (Table 4.2). The most productive agricultural soils of the continent are irrigated Entisols in the river valleys and the poorest are Entisols of the mountain massifs of the Sahara. Alfisols are the dominant soils in the Sudanian Zone south of the Sahara and in the semiarid fringe of southern Africa. Alfisols also are important soils on the semiarid western side of Madagascar. Arid-region Mollisols are inextensive and confined to the northwest coast. Soil associations in which Vertisols are dominant are few in number but Vertisols are important secondary soils, particularly along the south side of the Sahara.

Alfisols

Soils of this order are found between the Aridisols and humid-region soils, generally Ultisols. Insufficient data are available to separate the different kinds of Alfisols that undoubtedly are present in the areas delineated in Fig.4.1. Parent materials include sandy deposits as well as crystalline and ferromagnesium rocks.

A short summer rainy season of a few months duration is followed by a long dry season during which the grasses turn brown and die, leaving xerophytic trees such as acacias as the principal living plants. The rainy season becomes longer and the dry season shorter as the climate changes from arid to humid across the Alfisol zone. Leaching has been sufficient to develop an

TABLE 4.2

Area of dominant soil orders in arid regions of Africa

Soil order	Area (km^2)	Percent of	
		arid region	continent
Alfisol	2,085,000	11.8	7.0
Aridisol	4,890,000	27.7	16.4
Entisol	10,320,000	58.4	34.6
Mollisol	120,000	0.7	0.4
Vertisol	245,000	1.4	0.8
Total	17,660,000	100.0	59.2

argillic horizon. Carbonate dust blown in during the dry season probably accounts for the base saturation of more than 60% in most soils. Natric horizons apparently are rare in African Alfisols. Subsoils in the wetter areas sometimes are mottled with iron and manganese and may contain plinthite, the iron-rich material which hardens irreversibly to a brick-like substance upon exposure to the atmosphere. The topography is gently undulating, for the most part, and the vegetation is of the savannah and woodland types. Annual precipitation is between 500 and 900 mm except in West Africa where it may exceed 1,000 mm. Temperatures seldom go below zero.

Surface soils have a massive structure and become very hard when dry. They are, however, fairly friable when wet. The texture of the surface soil usually is sandy but the sands are compact rather than loose. There is no physical restriction to root development throughout the profile.

AL. Haplustalfs — Paleustalfs — Ustochrepts

Haplustalfs have thin brownish surface horizons overlying the argillic horizon. They have a low to moderate cation-exchange capacity, contain some weatherable minerals, and have kaolinite as the dominant clay mineral. Soil reaction varies from neutral to slightly acid in the surface and from slightly acid to moderately alkaline in the subsoil. Haplustalfs bordering on the Aridisol zone are calcareous in and below the argillic horizon. Organic-matter content is low. In the surface, the texture usually is sandy loam, whereas the subsoil is a sandy clay loam or sandy clay. Parent rock underlies the soil at 1—2 m.

Paleustalfs have characteristics similar to those of the Haplustalfs except that the surface soils are thicker, soil reaction is more acidic, there is more likelihood that a well-defined calcic horizon will occur in the argillic horizon, plinthite is present in many of the lower horizons, and soils are deeper over the parent material.

Ustochrepts are shallow (lithic) Inceptisols on the subhumid edge of the Alfisol zone. Inceptisols are weakly developed soils without argillic horizons. Ustochrepts have a light-colored surface horizon and an ustic moisture regime typical of the wet summer—dry winter equatorial regions. Most of them are noncalcareous in the surface but may be calcareous in the subsoil. The soils are shallow over igneous rocks. Vegetation consists of open woodland. Organic-matter content is low, base saturation is moderate to high, and the topography is gently undulating to sloping.

There probably are Tropaqualfs, Rhodustalfs, and Ultisols in the Alfisol zone, too.

Aridisols

Five associations of Aridisols are delineated on the soil map of arid Africa (Fig.4.1). Two of them are dominated by Calciorthids and Camborthids, and those two comprise most of the Aridisol areas. The one Salorthid association covers small areas, as does the Paleorthid association. There are inclusions of Entisols in most of the Aridisol areas and some Vertisols in the Saharan Aridisols.

AR1. Camborthids — Torripsamments — Haplargids

The largest continuous body of soils in this association is in the Sahelian Zone south of the Sahara, between the Sudanian Zone on the south and the Saharan Zone on the north. The remainder are mostly in the Sudan, Ethiopia, Somalia, and southern Africa. Crop production is hazardous because droughts are common. Sorghum and millets are the principal dryland crops, sorghum being the choice in the moister part of the area and millets in the drier part. The vegetative cover is sparse and consists of scattered xerophytic trees and shrubs with ephemeral bunch grasses and forbs. Mean annual precipitation is between 200 and 500 mm, in the main. The topography is level to gently undulating nearly everywhere, with low rocky ridges, sand dunes, depressions, and wadis providing variation in relief. In some places large plateaus project above the surrounding plain.

Camborthids are the Aridisols showing enough structural development to produce a cambic horizon but not enough for an argillic horizon. In Africa, Camborthids in this association generally have less than 1% organic matter in the surface, are noncalcareous in the upper 1—2 m of soil, have coarse to medium textures, sometimes are gravelly, and have a desert pavement 1—5 cm thick. Surface-soil color is gray to brown, whereas subsoils vary from yellow-brown to reddish brown. Soil depth ordinarily is from about 50 cm to 2 m over bedrock. The soils are nonsaline and nongypsiferous.

Torripsamments include barren and vegetated sand dunes and sandy soils having an undulating topography. Most of the small dunes and the undulating sands have sufficient bunch grass and shrub cover to hold them in place but on many of the overgrazed sandy soils plant cover has been destroyed and moving dunes have formed. The fixed sandy soils frequently have a surprisingly high organic-matter content in the top 1—5 cm of soil, sometimes exceeding 4%, which remains relatively high down to 1 m or more (Audry and Rossetti, 1962). Soil texture is generally fine sand throughout the profile, base saturation is high, and the soils are nonsaline, noncalcareous, and deep.

Haplargids occupy nearly level to gently sloping interdune areas and the older plateau and piedmont plain surfaces. They have a thin surface layer of brown to reddish-brown noncalcareous sandy loam or loamy sand, overlying a moderately thick noncalcareous argillic horizon of sandy clay loam. The lower part of the subsoil frequently is calcareous. Interdune Haplargids may be slightly to moderately saline below the argillic horizon. Most of the plateau and piedmont plain soils have a desert pavement. Organic-matter content is low. The soils are subject to severe wind and water erosion when they are cultivated.

Vertisols, Salorthids, Torriorthents, and Torrifluvents are other great groups in this association.

AR2. Calciorthids — Gypsiorthids

A part of northeast Somalia extending along the coast is in this association. The underlying rocks are gypsiferous and calcareous. Along the coast is a broad level plain 80—160 km wide; toward the interior the land slopes gently upward toward the mountains in the distance. Precipitation is 150 mm or less, temperatures are high, vegetation is sparse and consists mainly of xerophytic shrubs, and surface water is virtually nonexistent. The region can best be described as forbidding for man and animals. There is no good information about the soils.

Calciorthids in this area are intermingled with Gypsiorthids, but the former tend to predominate on the more sloping lands of the interior. Surface soils are calcareous sandy loams to loams several inches thick overlying strongly calcareous subsoils of about the same texture. Some lime concretions and crystals of gypsum are found in the lower part of the subsoil. Organic-matter content is low.

Gypsiorthids have a thin calcareous and gypsiferous sandy loam surface soil underlain by a calcareous and highly gypsiferous subsoil. Gypsum content is highest in the slight depressions in the coastal plain. The shrub cover is sparse and the organic-matter content is low everywhere.

Torripsamments and Salorthids are other great groups of soils.

AR3. Calciorthids — Torripsamments — Torriorthents

Soils of this association, located mostly in the eastern Sahara but also along the west coast and in southern Algeria, are a mixture of shallow and deep gravelly and stony soils, sandy soils, and clay plains. Little or no development is the common soil characteristic. The landscape is level, with some scattered low desert mountains breaking the monotony of the view across sandy plains having a moderate cover of xerophytic trees such as acacia or across regs completely devoid of any apparent vegetation except in the wadis. Clay plains are found only in the depressions of the undulating topography where sandy soils are dominant. There is some question whether the clay plains antedate or postdate the sand dunes around them, but it appears that the dunes have formed after the clay plains in most of western Sudan, Chad, and Niger.

Calciorthids are calcareous soils in which enough downward movement of lime has occurred to form a calcic horizon, usually within 30 cm of the surface. The texture is coarse — and frequently is gravelly — throughout the profile. Soils vary in depth to bedrock, usually within the range of 1—3 m. Organic-matter content is low, soils are gray to light brown in color, and salinity is low. Gypsum may be present in crystals and nodules in the lower subsoil but the amount is low. The topography is smooth, leading to the formation of regs, with their well-developed cover of desert pavement.

Torripsamments are sandy soils occurring largely as hummocks around the bushes or as low dunes in the western and southern Sahara part of this association and as long low dunes presenting an undulating topography in the eastern Sahara. Depending upon the clay content and the degree of soil development which has occurred, the interdune clay-plain soils are Haplargids or Pellusterts. The Torripsamments usually have a good-enough vegetative cover of shrubs and xerophytic trees to stabilize the sand and minimize wind erosion. Soils are deep, calcareous in the subsoil, and nonsaline.

Torriorthents are the shallow gravelly and stony soils over bedrock. They occur on the rocky ridges and desert mountains, on the stony plateaus called hamadas, and on the alluvial fans and piedmont plains. Vegetation is sparse and confined largely to shrubs and low trees except in the wadis where there may be thickets of trees such as *Acacia* and *Tamarix*. Soils on the gravelly alluvial fans and piedmont plains are usually deep and covered with a desert pavement of gravels and stones. On the hamadas, the soils are thin and covered with stones and large rocks. Most of the soils are calcareous to the surface and nonsaline but some Torriorthents in low-lying places are saline and may have a salt crust.

Other great groups represented in this association include Pellusterts, Haplargids, and Torrifluvents.

AR4. Paleorthids — Calciorthids — Salorthids

The one area of soils in this association is in the Hauts Plateaux of northern Algeria and Morocco between the Saharan Atlas and the coastal mountain ranges. The high plateaus are very large basins with gentle slopes toward the center where salt flats occur. In this part of Africa, the salt flats are called *chotts*, and several of them are tens of kilometers in length. Vegetative cover is largely grass and shrubs. The floor of the chotts is usually barren of vegetation and covered with a salt crust during the dry summers. A common sequence of soils begins with Salorthids in the bottom of the chotts, Calciorthids and Paleorthids in the level plain around them, Calciorthids on the gentle to moderate slopes above the plain, and Paleorthids on the terraces between the bottoms of the depressions and the mountains.

Paleorthids have a petrocalcic horizon within 1 m of the surface. The petrocalcic horizon is indurated enough to severely restrict root development but has many fractured sections through which water can move. Textures are dominantly loams and silt loams, the soils frequently contain calcareous nodules in the surface, organic-matter content is moderate, soil colors are brown and reddish brown, and the soils are nonsaline (Durand, 1959). The petrocalcic horizon commonly is within 50 cm of the surface and is only several centimeters in thickness, overlying a thick layer of calcareous nodules. Paleorthids derived from lacustrine deposits of the chotts usually contain considerable gypsum in the lime layer beneath the petrocalcic horizon whereas the Paleorthids on the terraces are free of gypsum. The petrocalcic horizon is distinguished by a very hard laminar section on the top.

Calciorthids resemble the Paleorthids but do not have a petrocalcic layer within 1 m of the surface. Instead, there is a calcic layer containing numerous calcareous nodules which, in time, may become indurated and form a petrocalcic horizon. Gypsum is usually present in the subsoils of the lower-lying Calciorthids of the chotts. The soils are deep, in the main.

Salorthids are found in the salt flat portion of the chotts where a high water table exists for several months of the year. There is a thick salt crust on the soil surface during the dry season. The water table fluctuates from near the surface to 1—3 m below the surface and is highly saline. Soils are medium to fine textured. Gypsum is present in high concentrations in most of the Salorthids.

Other great groups of some importance are Torrifluvents and Torriorthents.

AR5. Salorthids — Gypsiorthids — Torriorthents

The arid coastal plain of the Red Sea is the site of these soils. The delineation is duplicated on the Arabian peninsula side of the Red Sea. In both cases, a chain of desert mountains parallels the coastal plain. Next to the sea is a string of saline flats having a high water table and scattered sand dunes. Inland of the saline flats is a broad, level to gently undulating sandy plain which merges into alluvial fans along the front of the mountains. Vegetation is sparse or absent on the saline flats but the plant cover is fair to moderate on the sandy plain where *Paspalum* and *Panicum* are the dominant grasses and *Acacia* is the principal tree. Various shrubs provide a thin cover on the coarse-textured alluvial fans. All of the soils are weakly developed, at best, organic-matter content is low, salinity is widespread along the coast, and the soils are calcareous. At Port Sudan, the annual precipitation is 94 mm and the annual temperature is 28°C, with a maximum of 47°C and a minimum of 10°C.

Salorthids are concentrated along the shore where the water table is 1 or 2 m below the surface. Salinity is high throughout the profile. Surface soils have a coarse to medium texture, subsoils are medium to fine textured, gypsum commonly is found in the subsoil, and small sand dunes are not unusual.

Gypsiorthids are found on the seaward side of the sandy plain where the soils are finer textured than they are on the inland side. The Gypsiorthids have a gypsic horizon at or near the surface, are calcareous and at least moderately saline, and are moderately deep to deep, in general.

Torriorthents occupy the alluvial fans and the inland side of the sandy plain. The alluvial fans are gravelly and stony and have a desert pavement as well as some small hummocks of sand around bushes. The soils are slightly to moderately calcareous and generally are deep, with little or no evidence of development.

Calciorthids and Torrifluvents also are present in this association.

Entisols

E1. Torripsamments — Haplargids

The Sahara is notable for the great sand seas (ergs) which stretch for hundreds of kilometers across the desert in a seemingly endless succession of massive dunes. Outside of the Sahara, the only really large area of high dunes is in the southern part of the Namib Desert along the coast of South West Africa. Sand dunes of all sizes and shapes are common in most of the Aridisol and Entisol areas of Africa but the sand seas represent a distinctive landscape. They consist of dunes rising to 100 m or more above the interdune areas, completely bare of vegetation, easily erodible by wind, and showing no profile development. When the dunes are close together, the interdune areas consist of deep sands; when the dunes are farther apart, the interdune soils may have a cover of sand or they may have an argillic horizon and qualify as Haplargids. The Grand Erg Oriental and Grand Erg Occidental in Algeria are representative of the sand seas having a continuous cover of sand in and between the dunes.

Torripsamments are the sandy Entisols and, in the Saharan ergs, are very deep. Whatever development may have occurred in them during more humid times has been obliterated as the result of the mixing caused by wind erosion. The soils are noncalcareous or only slightly calcareous in the surface 1 or 2 m. Organic-matter content is very low because the sands are nearly devoid of living matter, including microorganisms.

Haplargids are Aridisols with an argillic horizon, usually within 30 cm of the soil surface. They are found, in this association, in the level interdune areas which vary greatly in size depending upon how far the dunes are separated from one another. Haplargids have a thin surface layer of sandy loam overlying a moderately thick sandy clay loam argillic horizon which may be calcareous in the upper part and usually is calcareous in the lower part. In many places, the argillic horizon is exposed on the surface as the result of wind erosion. Vegetative cover is very sparse or nonexistent, and organic-matter content of the soil is low. The soils are nonsaline.

There are some Torrifluvents and Torriorthents in this association.

E2. Torripsamments — Paleorthids

The Kalahari Desert of Botswana, South West Africa, and South Africa constitutes the one area where this soil association occurs. It is a level to gently undulating featureless plateau, at an altitude of 700—1,500 m, covered with sand and, on the borders, with exposed petrocalcic horizons. The sand, in most cases, is about 12—60 m thick over granitic and quartzose rocks. Sand dunes stabilized with coarse tuft grasses and shrubs rise 10—15 m above the level of the sandy plain. The dunes decrease in average height from southwest to northeast. Surface-water supplies in the center of the desert are virtually nonexistent except where water stands in pans (playas) immediately following rains. Several dry watercourses cross the desert in the south and one perennial stream (the Okavango River) does the same in the north.

Despite its name, the Kalahari Desert has a fair to good vegetative cover due to a mean annual rainfall ranging from 150 mm in the southwest to 500 mm in the northeast and to sandy soils which absorb much of the precipitation. Plant cover is grass (*Aristida*, *Eragrostis*, *Andropogon*) with a fairly dense growth of low bushes and shrubs in the south (*Terminalia* and *Rhigozum*) and bushes and trees (*Capaifera*, *Commiphora*, *Adansonia*, and *Acacia*) in the north. Soil erosion is only of minor importance.

Torripsamments are, in the Kalahari, deep aeolian sands occupying the central part of the desert. There is only slight variation in texture with soil depth. Soil color varies from reddish brown to brick red in the surface 1—2 m to yellowish brown in the substratum. The topography is undulating, with many low dunes. Interdune areas have a little more clay than in the dunes and the soils are more compact, but slowly permeable soils are confined to the topographic depressions called pans. The sands are stabilized by vegetation.

Paleorthids on the outskirts of the Kalahari Desert have a petrocalcic (indurated calcium carbonate) horizon underlying Kalahari sands and exposed on the surface in many places. Technically, only those sands having a petrocalcic horizon within 1 m of the soil surface are Paleorthids. Most of the sands have the petrocalcic horizon below 1 m but it can be found at any depth between the surface and 4 m. As with the Torripsamments, soil color is reddish brown to brick red and subsoil color above the petrocalcic layer is yellowish brown. There are occasional iron-oxide concretions in the subsoil and upper part of the petrocalcic horizon. Van der Merwe (1962) attributes the formation of the petrocalcic horizon to the dissolution of lime in the underlying rocks, followed by upward movement and precipitation of the dissolved calcium carbonate in the overlying sand. The creamy white petrocalcic layer is 30—60 cm thick, fissured sufficiently to permit easy root and water penetration through it, and has an extremely hard crust about 3 mm thick. Beneath this horizon is a highly calcareous softer layer containing nodular and powdery calcium carbonate which may be several meters thick.

Other great groups in the Kalahari include Torrifluvents, Torriorthids, and Haplargids.

E3. Torriorthents — Calciorthids

Regs and hamadas are the typical surfaces on which Torriorthents and Calciorthids have developed. Regs are plains covered with a desert pavement of gravels and stones; they are also known as stony deserts. The underlying soil could be nearly anything but usually is coarse textured and moderately deep to deep. Regs are classified as residual and transported. Residual regs occupy plateau positions, usually, and the stones on the surface are produced by the mainly physical weathering of the parent consolidated rock and the removal by wind of fine material originally present between the stones. Transported regs are found typically on alluvial fans and terraces where gravels have been deposited following floods in the higher-lying lands. Residual regs in North Africa and southwest Asia are called hamadas when they occur on plateaus. Regs are also known as serirs in northeast Africa, gobis in China and Mongolia, and gibber or stony plains in Australia. Organic-matter content is low and the soils are generally nonsaline but some hamada soils near the coast may be saline.

Torriorthents are either shallow rocky soils belonging to the subgroup of Lithic Torriorthents or are the variable-depth gravelly soils of moderate slopes which represent the most prevalent Torriorthents of the arid regions. Lithic Torriorthents of the hamadas and residual regs have a stony sandy loam to loam surface soil of a few centimeters in thickness underneath the stony desert pavement. Beneath that is a somewhat finer textured and stony horizon 10—20 cm thick which grades into weathered rock. In addition to the angular rock fragments which are common on the surface, bedrock outcrops occasionally on the surface. Typical Torriorthents are transported regs on moderately sloping alluvial fans at the base of mountains and hills. They have a gravelly sandy loam surface and subsoil, with calcium carbonate accumulations beginning at 20—30 cm. Bedrock lies at variable depths beneath the gravelly soil.

Calciorthids are found on the gentle slopes of piedmont plains, alluvial fans, and stream terraces. They are included in the transported reg category when they have a desert pavement, as they usually do in the Sahara. Calciorthid profiles resemble those of the typical Torriorthents in being gravelly sandy loam in texture but they differ in having a calcic horizon at depths of 10—20 cm. Most of them are 1 m or more in depth.

Rock outcrops consisting of bare rock are widespread in this association. Other great groups include Torrifluvents, Paleargids, and Haplargids.

E4. Torriorthents — Calciorthids — Paleorthids

Soils in this association are dominantly shallow to very shallow over a variety of rocks and petrocalcic horizons, with some deeper soils in which a calcic horizon is present. The topography ranges from level to steeply sloping and the landscape is dissected by numerous dry watercourses which carry water only after rains. Water erosion has left a deep imprint on the land but small and large sand dunes are common, as well, in the more level areas. Vegetation is sparse except in southern Africa where there is a moderate cover of grasses, shrubs, and trees.

Most of the *Torriorthents* belong to the Lithic subgroup in which bedrock usually is at a depth of 10—50 cm but outcrops frequently on the surface, especially on the steeper slopes. Surface soils may be of any texture from sandy loam to silty clay loam, with moderately coarse and medium textures predominating. The great majority of the soils are slightly calcareous unless the parent rock is limestone, in which case the soils may be highly calcareous immediately above the limestone. Other Torriorthents are moderately deep and deep gravelly sandy loams on alluvial fans and terraces. They are slightly to moderately calcareous and are low in organic matter.

Calciorthids are the gravelly sandy loam Torriorthents on alluvial fans, piedmont plants, and stream terraces in which a calcic horizon has developed within 1 m of the surface. Deep gravelly soils in this association grade from Torriorthents having calcium carbonate — if present — distributed more or less uniformly throughout the profile to Calciorthids having a horizon in which calcium carbonate has accumulated but is not indurated to Paleorthids having an indurated calcic (petrocalcic) horizon. Calciorthids are extensive in Somalia and Ethiopia.

Paleorthids, the Calciorthids with an indurated horizon of calcium carbonate, are common in South Africa and South West Africa and in North Africa from Morocco to Egypt. In North Africa, there are soils with petrogypsic horizons as well as petrocalcic horizons, especially in Algeria and Tunisia. Durand (1959) refers to petrogypsic layers as "encroûtement gypseux"; petrocalcic formations are "encroûtement calcaire".

Torripsamments, Torrifluvents, and Gypsiorthids are among the other great groups in this association. There is a large area of stabilized sandy soils (Torripsamments) overlying clay plains west of Khartoum in the Sudan. The area in which they occur is called the Qoz and is a savannah with low *Acacia* and *Adansonia* trees.

E5. Torriorthents — Rock Outcrops

Desert hills and mountains are the dominant landscape features in this association. Soils are undeveloped or weakly developed, generally coarse textured except in depressions or in the extensive clay plains in the Sudan and Chad, mostly shallow with rock outcrops in the more sloping areas, low in organic matter, and slightly to moderately calcareous. Vegetation is sparse and consists largely of shrubs over most of the area but the plant cover is fairly good at higher elevations in the massifs of the central Sahara (Hoggar, Tibesti, Air) where the precipitation is relatively high. Water erosion has been the principal agent of landscape formation due to the torrential character of the rain storms and the shallowness of the soils.

Torriorthents and *Rock Outcrops* (a land type rather than one of the soil orders) are intimately intermingled in the mountains, ridges, and hills. Barren rock and shallow soils can be found nearly everywhere in this association but there are significant amounts of deep gravelly soils on alluvial fans and piedmont slopes and deep moderately fine to fine textured soils of the clay plains. All except the bare rocks are variants of Torriorthents. Salinity levels are low in the clay plains as well as elsewhere. Most of the soils are calcareous in some part of the profile.

Torrifluvents represent the principal other great group but there also are Haplargids, Torripsamments, and Calciorthids.

E6. Torrifluvents — Salorthids — Fluvaquents

Recent alluvial soils of irrigated valleys, deltas, depressions, and swampy areas comprise most of this association. Salinity is a present or potential problem nearly everywhere and high water tables are common. In addition to the delineated areas, many Saharan oases have this same grouping of soils, where the continuing hazard for crop production is the combination of high water tables and saline soils. There are three large irrigated valleys in this association (Nile, Senegal, Niger), an extensive swamp (Okavango in Botswana), a deep saline depression (Qattara in Egypt), and a shallow one (Makarikari in Botswana), and an irrigated alluvial plain in Morocco. In Africa, Torrifluvents and Salorthids are intimately mixed, in the main, and differ only in the presence or absence of salt accumulations and a high water table.

Torrifluvents are soils formed in recent alluvial deposits along waterways and deltas. They usually are stratified with soil layers of differing textures and the organic-matter content varies in an irregular fashion with depth of soil. Surface-soil texture may be anything from sand to clay but commonly is a loam, silt loam, or clay loam in the broad valleys and deltas. Soil reaction generally is slightly alkaline to slightly acid, soils frequently are somewhat saline, and veins of calcium carbonate and gypsum may be present in the subsoil. The soils typically are deep. When irrigated, the medium to fine textured soils can be highly productive, as has been demonstrated in the Nile, Senegal, and Niger valleys.

Salorthids are saline soils having a water table close enough to the surface during at least part of the year to permit salts to move upward and be deposited at or near the surface. In this association, Salorthids are recent or old alluvial soils having a salic horizon and an intermittently high water table. They are prevalent in the lower Nile and Senegal valleys, the Qattara depression, and the Makarikari depression.

Fluvaquents are Entisols with a high water table for part of the year, slight to moderate evidence of stratification, and a relatively high content of organic matter. These are the soils of the deltas and the Okavango swamp.

Mollisols

M. Haploxerolls — Ustorthents — Rhodoxeralfs

Mollisols and their associated Orthents and Alfisols are restricted to the moister regions along the north coast of Africa where precipitation effectiveness is high due to the Mediterranean climate of wet winters and dry summers. Haploxerolls and Rhodoxeralfs are found together throughout the delineation and grade into one another. Their profile characteristics and chemical properties are similar but most of them appear to resemble Mollisols more than they do Alfisols. Ustorthents probably occupy more of the area than do the Mollisols and Alfisols because most of the land is moderately to steeply sloping and soils are shallow. The vegetative cover is mostly trees on the more sloping land and grasses and shrubs in the lowlands and on the river terraces.

Haploxerolls are Mollisols of Mediterranean climates. The organic-matter content of the mollic epipedon may be as much as 3%, surface-soil texture is loam to clay loam, and the clay content of the subsoil is a little higher than in the surface. Soil pH is 7 or slightly more and the base saturation is more than 80%. While the surface soil sometimes is free of carbonates, there often is an accumulation of calcium carbonate in the lower part of the profile. The native vegetation is grasses with scattered shrubs. The topography is level to gently sloping in the broad valleys where these soils commonly occur.

Ustorthents are the shallow soils on the tops and sides of hills and mountains. They show little evidence of development and are underlain by partially weathered rock at 10—30 cm. Soil texture varies from sandy loam to clay loam. The vegetation consists of trees and shrubs, with some grasses present in the more open spaces.

Rhodoxeralfs are the Xeralfs with a red argillic horizon. Base saturation of the B (argillic) horizon is usually 60% or more and increases with depth, in most cases. The pH is near neutrality, the A and B horizons are free of carbonates, and the argillic horizon has a blocky or prismatic structure. The organic-matter content of the surface soil is about 1% and it becomes somewhat less in the subsoil. Vegetative cover consists of an open woodland with an understory of shrubs and some grasses. Slopes are moderate to steep.

Among the other great groups in this association are Rendolls, Argixerolls, Torrifluvents, and Salorthids.

Vertisols

V1. Chromusterts

The one place where this great group occurs is in the clay plain between the Blue Nile and the White Nile in the Sudan. Smaller — not delineated — areas of the same kind of soil lie east of the Blue Nile. Vertisols occupy nearly 100% of the land area in the triangle between the two Nile branches. Sand dunes are found occasionally on top of the Vertisols in places where sand has blown in from stream channels, or from sandy lands adjoining the plain. The well-known Gezira Scheme which was established in 1925 to produce long-staple cotton under irrigation is located on the Chromusterts of the clay plain, as is the Managil extension to the Gezira Scheme. The entire area is flat. Irrigation water is taken from the Blue Nile at Sennar and Roseires dams. The native vegetation is savannah grasses and *Acacia*. Parent material is Blue

Nile alluvium from basaltic rocks in Ethiopia underlain by White Nile alluvium laid down when the area was a lake.

Chromusterts of the Sudan clay plain average about 10 m deep over sand. Locally, the soil is called "Gezira clay" and is a classic example of the cracking clays known as Vertisols. Gezira clay consists of dark-brown clay to a depth of 60—90 cm, followed by a dark grayish-brown clay down to 120—130 cm which is underlain by a brown to yellowish-brown clay extending to a depth of 10 m or more. Underneath the clay is a thick deposit of sand. Soil structure is angular blocky, the soils have a high degree of plasticity, the clays are montmorillonitic, carbonate nodules are common throughout the profile, there is a gypsum accumulation layer starting at approximately 1 m, and the soils are nonsaline or slightly saline in the surface and saline and sodic (more than 15% exchangeable sodium) in the subsoil. Lastly, as for all Vertisols, the soil has deep (up to 1.4 m) and wide (1—12 cm) cracks when dry. The brown surface soil "tongues" into the gray layer wherever cracks extend into the latter. Surface soil falling or being washed into the cracks leads to the churning which is typical of cracking clays. Churning develops when the cracks become filled with soil from above and the soil between the cracks is rewetted and expands. Since the cracks now have soil in them, expansion leads to upward displacement of the soil mass in order to relieve the pressure. A gilgai (hummocky) relief ordinarily would develop from the churning action, but it is obscured by cultivation in the Gezira (Finck, 1961). In this slowly permeable soil, cracks are an important avenue for water penetration into the subsoil.

On the west side of the clay plain, east of the White Nile, the soils apparently are more saline and contain more exchangeable sodium than do those of the Gezira Scheme on the east side of the plain. They are said to be less productive than the Gezira soil.

V2. Chromusterts — Pellusterts — Haplustalfs

This association occurs only in southern Chad east and southeast of Lake Chad. The topography is level to gently sloping to undulating, with broad depressions between low hills. Vertisols are found in the depressions, Alfisols on the slopes and tops of hills. The vegetation is savannah on the uplands and a heavy cover of trees in the depressions. Of the principal trees, *Acacia* is found in both places and *Guiera* on the upland soils. Precipitation across the area varies from about 500 to 900 mm, falling between June and October with a maximum in August. A typical cross-section of the landscape shows sandy Alfisols on ancient sand dunes and low hills, brown and yellow-brown Chromusterts with calcareous nodules in the subsoil on the slightly sloping sides of the depressions and on stream terraces, and black Pellusterts in the bottom of the depressions.

Chromusterts and *Pellusterts* differ, in this region, in color of the surface soil, in the prevalence of carbonate nodules in the subsoil, and in the amount of clay and organic matter they contain. Chromusterts on the gentle slopes or slightly elevated places have a brown color, fair drainage, many carbonate nodules, about 0.5—1% organic matter in the surface 15 cm of soil, and 40—55% clay. Pellusterts in the depressions are black in color and poorly drained, have few carbonate nodules, contain 0.8—1.4% organic matter in the surface, and have 50—70% clay. The clay is montmorrillonitic and cracks deeply during the dry season. Base saturation varies from 80 to 100% and the pH is usually close to neutrality or is on the alkaline side.

Haplustalfs are leached sandy soils of the uplands. Typically, they have a brown to red-brown sandy loam surface soil underlain at 30 cm by a thick sandy clay loam argillic horizon. Base saturation in the surface soil is 50% or more, but is less than 40% in the deep subsoil. They are moderately to strongly acid, low in organic matter, noncalcareous at any depth, low in soluble salts, and have cation-exchange capacities of 1—3 mequiv. per 100 g of soil.

REFERENCES

Audry, P. and Rosetti, C., 1962. *Observations sur les sols et la végétation en Mauritanie du sud-est et sur la bordure du Mali (1959 et 1961). Prospection Ecologique. Etudes en Afrique Occidentale.* FAO, Rome, 267 pp.

D'Hoore, J.L., 1964. *Soil Map of Africa. Scale 1 to 5,000,000. Explanatory Monograph and Map.* Commission for Technical Cooperation in Africa, Lagos, Joint Project No. 11, Publication No. 93: 205 pp. + 13 maps.

Durand, J.H., 1959. *Les Sols Rouges et les Croutes en Algérie.* Direction de l'Hydraulique et de l'Equipment Rural, Service des Etudes Scientifiques, Alger, Etude Générale No. 7. 188 pp.

Dyer, R.A., 1955. Angola, South-West Africa, Bechuanaland and the Union of South Africa. In: UNESCO, *Plant Ecology. Reviews of Research. Arid Zone Res.*, VI: 195—218.

Finck, A., 1961. Classification of Gezira clay soil. *Soil Sci.*, 92: 263—267.

Meigs, P., 1953. World distribution of arid and semi-arid homoclimates. In: UNESCO, *Reviews of Research on Arid Zone Hydrology. Arid Zone Res.*, I: 203—210.

U.S. Department of Commerce, 1969. *Climates of the World.* Environmental Data Service, Environmental Science Services Administration, U.S. Department of Commerce, Washington, D.C., 28 pp.

Van der Merwe, C.R., 1962. *Soil Groups and Subgroups of South Africa. Dep. Agric. Tech. Serv. Repub. S. Afr., Sci. Bull.*, No. 356: 355 pp.

Chapter 5

ASIA

INTRODUCTION

The arid regions of Asia (Fig.5.1) are dominated by Aridisols and Entisols with significant areas of Mollisols and Vertisols (Fig.5.2). There are no large areas of Alfisols although some have been recognized in the hilly region southeast of the Rajasthan (Thar) Desert and in the Gangetic Plain in India (Raychaudhari and Govinda Rajan, 1972). The principal single source of information for the soil map of the arid regions of Asia was the map prepared under the direction of Kovda and Lobova (1971), with assistance from soil scientists of the U.S.S.R., China, India, and other Asiatic countries. Detailed information on soils was available for the Soviet Union, Mongolia, Iran, and Iraq. Less-detailed data were available for India and Israel, and only sketchy information was obtained for the other countries.

ENVIRONMENT

Climate

Asia is a vast continent, and the climatic variation is great (Table 5.1). In the countries extending from the Mediterranean Sea to Soviet Central Asia the climate is typically Mediterranean, with wet winters and dry summers. In the U.S.S.R., there is a gradual change to the continental climate which prevails over the northern part of Soviet Central Asia and in China and Mongolia. In the latter areas, winters are generally rather dry and the summers are moist. South of the Mediterranean, in the Arabian peninsula, rainfall is scanty throughout the year except in the southern mountains, which feel the influence of the summer monsoon. The monsoon climate dominates Pakistan and India.

Throughout most of the arid portion of Asia (Fig.5.1), precipitation averages less than 250 mm except in places near the mountains, in central India, and in the Mollisol belt which traverses the northern edge of the arid regions. There the precipitation increases to between 250 mm and 500 mm and even exceeds 500 mm on the Deccan Plateau of India where Vertisols are extensive.

Winters are long and cold in the north, becoming shorter and less severe toward the south. Summers are hot everywhere. The maximum temperature

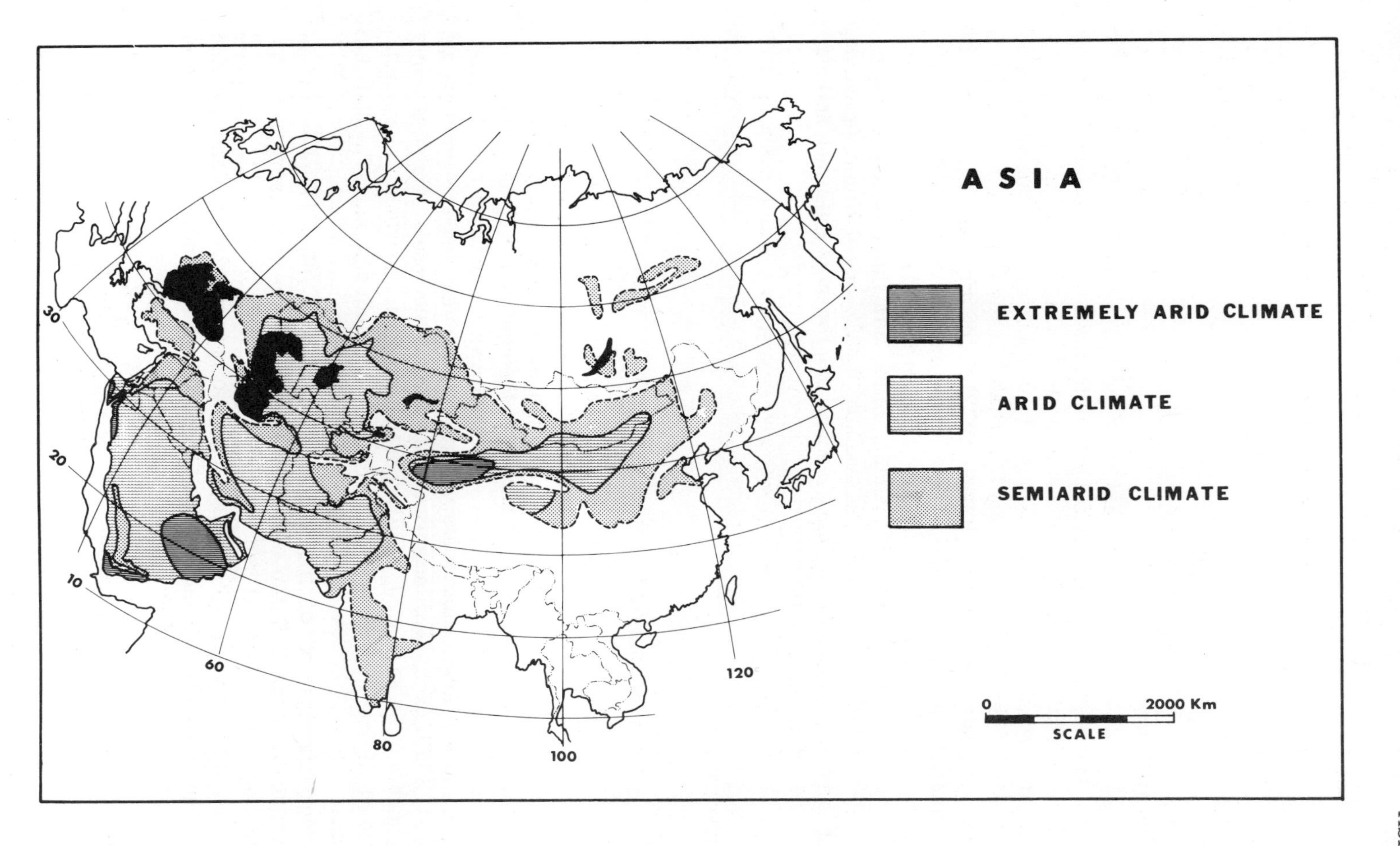

Fig.5.1. Arid regions of Asia (after Meigs, 1953).

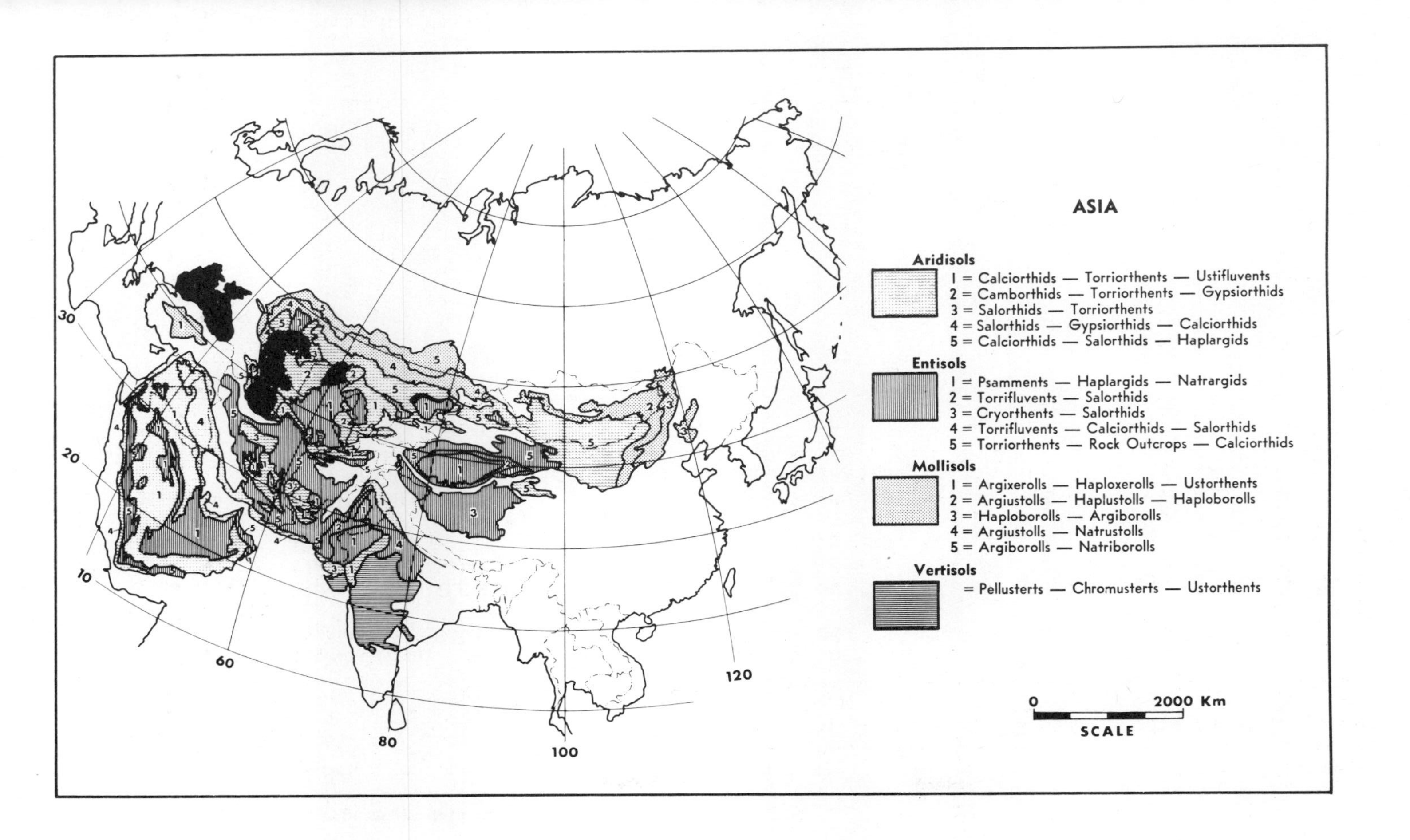

Fig.5.2. Arid-region soils of Asia. Polar soils not shown.

TABLE 5.1

Precipitation and temperature data for weather stations in Asia

Type of climate	Station	Elevation (m)	Precipitation (mm)													Average annual temperature (°C)
			J	F	M	A	M	J	J	A	S	O	N	D	annual	
Mediterranean	Baghdad (Iraq)	34	23	25	28	13	2	0	0	0	0	2	20	25	138	22
	Tashkent (U.S.S.R.)	478	53	28	66	58	36	13	5	2	2	30	38	41	372	12
Continental	Semipalatinsk (U.S.S.R.)	216	23	13	13	15	30	38	28	33	18	30	28	25	294	2
	Ulan Bator (Mongolia)	1307	0	0	2	5	8	25	74	48	20	5	5	2	194	−3
	Kashgar (China)	1310	15	2	13	5	8	5	10	8	2	2	5	8	83	12
Monsoon	Karachi (Pakistan)	4	13	10	8	2	2	18	81	41	13	2	2	5	197	26
	Hyderabad (India)	530	8	10	13	30	28	112	152	135	165	64	28	8	753	27

Source: U.S. Department of Commerce (1969).

recorded at Baghdad (Iraq) was 49°C and the minimum was −8°C, whereas at Ulan Bator in Mongolia the maximum was 36°C and the minimum was −44°C. Extremes were not so great at Karachi (Pakistan) or Hyderabad (India), the former located on the Arabian Sea and the latter on the Deccan Plateau. The maximum and minimum temperatures at Karachi were 48°C and 4°C, respectively, while at Hyderabad they were 44°C and 6°C. Precipitation and temperature data for a few weather stations located in the different climatic zones of arid Asia are given in Table 5.1. Temperatures at Ulan Bator are so low that the area verges upon a cold rather than a hot desert; so does Tibet, apparently, although no reliable temperature records were available for locations there.

Geology and geomorphology

Arid Asia is a great plain broken by mountain ranges and marked by large and small depressions (closed basins) having no surface outflow to the sea. Surface deposits are diverse, ranging from stony, gravelly, and sandy to clayey in texture. As is typical of arid regions, the uplands are coarse textured and the lowlands are fine textured.

Various terms have been used to describe surficial conditions in arid Asia. Rock deserts and stone deserts include two types of surfaces, both of which are covered with large stones, rocks, and boulders and have little vegetation: hamadas are more or less smooth uplands; rough broken land is steeply undulating, hilly, or mountainous land. Clay deserts and clay plains are extensive fine-textured soils occupying the closed basins so typical of the arid regions. Most of the clays probably are slightly or moderately saline but without a surface crust of salt. Salt deserts and saline plains usually are similar to the clay plains and clay deserts in the texture of the underlying soil and in their size, but have salt crusts on the surface and may have a high water table from which salt rises to the surface. Sand deserts and sandy deserts may refer to extensive sand seas (ergs) with giant dunes, such as the Rub' al Khali, or to broad regions of moderately sized sand dunes having fine-textured inter-dune areas, such as the Kara Kum and Kyzyl Kum sands.

Gravel deserts are plains of coarse-textured soils covered with the pebbles, gravels, and small stones which constitute a desert pavement and are effective barriers to water and wind erosion. They are the gobis of Central Asia and the regs of North Africa. Steppes are timberless, level to gently undulating semiarid regions with a fairly continuous vegetative cover of xerophilous shrubs and herbs. They are wetter than what is usually called deserts and drier than the woodlands and savannahs.

Salt accumulations are common in the closed basins. The largest salt desert in the world is the ancient alluvial plain called the Dasht-i-Kavir (Iran), in which runoff water from surrounding higher lands has evaporated, leaving behind salts dissolved in the water. The highlands of Tibet, much of Mongolia, and the northern part of Soviet Central Asia are spotted with saline depressions. The Central Asia depressions, which become intermittent lakes following rains, are usually smaller than the closed basins in the Iranian Plateau and in the Arabian peninsula.

Wind erosion is the dominant erosional feature in places where the surface soil is sandy, but water erosion controls landscape development in the stony and gravelly upland soils. Sand seas — called *ergs* in North Africa — occupy two large areas: the Takla Makan Desert in Sinkiang and the Rub' al Khali (Empty Quarter) in Saudi Arabia. There, sand dunes follow one after another for tens or hundreds of kilometers, some of them reaching as high as 100 m or more. Other large dune fields can be found in Soviet Central Asia (Kara Kum and Kyzyl Kum sands), eastern Iran (Dasht-i-Lut), central and northern Saudi Arabia (Ad Dahna and An Nafud) southern Afghanistan (Registan), northern Pakistan (between the Jhelum and Indus rivers), and along the south and central borders between Pakistan and India.

Surficial gypsum deposits are extensive and thick in the plains of Iraq and Syria and in the Ust-Urt Plateau in the U.S.S.R. Gypsum is a major constituent of soils in the lowlands of Iran and along the coast of the Persian (Arabian) Gulf and the Red Sea. Cyclic (sea-borne) salts are responsible for much of the salinity in soils bordering the Gulf and the Red and Arabian seas.

Major permanent rivers, arising in the mountains and crossing the arid plains to empty into the seas, include the Tigris and Euphrates in Iraq; the Amu Darya, Syr Darya, and Volga in the Soviet Union; and the Indus in Pakistan. Numerous other rivers carrying large volumes of water, such as the Helmand in Afghanistan and the Tarim in Sinkiang, end up in closed basins or simply disappear in the sands.

Vegetation

Shrubs are dominant throughout the drier parts of arid Asia; only in the Mollisol belt in the north and east are grasses dominant. There is less herbaceous vegetation in the Arabian peninsula than in the remainder of the region.

A psammophytic shrub, *Calligonum*, can be found on sand dunes from the Gobi Desert to the Rub' al Khali in Saudi Arabia (McGinnies, 1968). Four other shrubs which are widely distributed on sandy soils are *Haloxylon*, *Ephedra*, *Artemisia*, and *Salsola*. The principal grass in those areas is *Aristida*. On the uplands of alluvial fans and piedmont plains, shrubs such as *Anabasis*, *Artemisia*, and *Astragulus* are common in the Arabian, Iranian, and Soviet Central Asia deserts; in the colder deserts of eastern Central Asia, *Calligonum*, *Salsola*, and *Haloxylon* are found on the deeper coarse-textured soils. On the pebble-or-stone-covered surfaces called *hamadas* in the Arabian Desert and *gobis* in China and Mongolia, *Zygophyllum*, *Anabasis*, and *Ephedra* shrubs occur across the entire region but *Nitraria* is centered in the Gobi Desert,

Alhagi in Soviet Central Asia, and *Euphorbia* in Pakistan and India. *Stipa* grasses are common in the colder deserts, *Aristida* and *Poa* in the warmer ones.

Shrubs found on saline soils include *Suaeda*, *Haloxylon*, *Seidlitzia*, and *Salsola*, with *Halogeton* locally important in the Takla-Makan Desert of Sinkiang. *Tamarix* trees grow along watercourses everywhere and on some saline flats in the warmer regions, where the water table is not too deep. Unstable sand dunes, the great salt desert of Dasht-i-Kavir and the takyrs of Central Asia are virtually devoid of higher plants. Takyrs, the broad depressions having slowly permeable soils, support an algae—lichen community in the Soviet Union and, probably, elsewhere.

SOILS

Aridisols, closely followed by Entisols, dominate the soil associations of arid Asia and adjoining areas in the European part of the U.S.S.R. bordering the Caspian Sea (Table 5.2). In the table, the Asian continent is considered to extend from Turkey to Japan and to include the Soviet republics east of the Caspian Sea and the Russian Federated Republic east of the Ural Mountains.

A notable feature of Asian arid-region soils is the large block of Vertisols (cracking clays) in India. The 780,000 km^2 of Vertisols there is slightly more than are delineated in Australia and considerably more than in the other continents. There arc no associations dominated by Alfisols in Fig.5.2 but

TABLE 5.2

Area of dominant soil orders of arid regions of Asia*

Soil order	Area (km^2)	Percent of	
		arid region	continent
Aridisol	5,920,000	41.1	13.6
Entisol	4,855,000	33.7	11.1
Mollisol	2,850,000	19.8	6.5
Vertisol	780,000	5.4	1.8
Total	14,405,000	100.0	33.0

*Does not include arid-region soils of European part of U.S.S.R. as delineated in Fig.4.1, which amount to 70,000 km^2 of Aridisols, 95,000 km^2 of Entisols, and 225,000 km^2 of Mollisols.

there are some Alfisols present, principally in India. Significant areas of arid-region Mollisols are confined to mid-latitudes in the heart of Asia and to the central plain of Turkey.

Aridisols

Five kinds of Aridisols and associated soils have been delineated on the soil map of Asia. Two of them have Calciorthids as the dominant soil group, in two of them Salorthids are dominant, and Camborthids are the major soil group in the fifth. The associated order is Entisols. There are inclusions of Mollisols in the northern perimeter of the Aridisols.

AR1. Calciorthids — Torriorthents — Torrifluvents

Of the five Aridisol areas on the map, this one is the most extensive and most widely dispersed over the continent. Large bodies of the association are found in Pakistan and India, in Soviet Central Asia, in Iran and in the countries of the Arabian Desert. The landscape is a combination of desert mountains, alluvial fans, piedmont plains, and flood plains of intermittent and permanent streams. The vegetation is a sparse cover of xerophytic shrubs and some grasses, with *Tamarix* growing along the permanent streams and the better-watered intermittent streams. Sand dunes are common features in the more level areas. Calciorthids are Orthids having an ochric epipedon and a calcic horizon within 50 cm of the surface. The surface soil commonly is calcareous. They do not have a duripan or a petrocalcic (indurated calcium carbonate) horizon within 1 m of the soil surface. They usually are dry unless irrigated. Parent material for the soils is highly calcareous or large amounts of lime have been added in dust falling on the soil. There may be a salic horizon underlying the calcic horizon. There is no argillic horizon. Organic-matter content is low; permeability to water is good unless a duripan or a petrocalcic horizon is present; surface-soil color is mainly gray or gray-brown.

In the Russian terminology, soils having Calciorthid properties are included among the Serozems. Serozem is a word that has meant many things to many people ever since S.S. Neustruev introduced the term in 1908 (Lobova, 1960). The soils that Rozanov (1951) called Serozems ran the gamut from undeveloped soils (primitive Serozems) to moderately developed soils (kyr or structural Serozems) on old landscapes, and included dark-colored calcareous soils which would qualify as Mollisols in the U.S. Comprehensive System. Primitive Serozems now are called desertic soils. Kyr Serozems have an argillic horizon and would, therefore, belong to the Argid suborder rather than the Orthid. In some of the other Asiatic countries where Serozems have been mapped the broad classification of Rozanov seems to be used rather than the more restrictive present-day Russian classification.

Rozanov (1951) describes an "ordinary" Serozem (Calciorthid) from the Turkmen S.S.R. as given in Table 5.3. An accompanying description notes

TABLE 5.3

Description of an "ordinary" Serozem (Calciorthid) from the Turkmen S.S.R. (After Rozanov, 1951)

Horizon	Depth (cm)	Description
A1	0—11 cm	light grayish with straw-yellow tinge; light loam; platy-laminor down to 4 cm; unstable lumpy structure below 4 cm
A2	11—20 cm	lighter-colored than the A1; weakly lumpy light loam; many roots
B1	20—47 cm	straw-yellow with grayish tinge; loam with loose structure
B2	47—70 cm	straw-yellow; porous; loess-like loam; root holes of burrowing animals; many "white eyes" of calcium carbonate; carbonate concretions up to 3 cm in diameter
C1	70—115 cm	light straw-yellow; loess-like light loam; carbonate deposits down to 80—95 cm; gypsum below 80—95 cm
C2	115—160 cm	whitish straw-yellow; fine-silty; loess-like light loam; slightly porous; abundant gypsum in fine crystals, increasing with depth

that the calcium carbonate content of the A horizon varies between 4 and 16% and that of the B2 from 14 to 22%. Soluble salt in the horizons overlying the gypsum horizon is low. In Rozanov's description, the B2 is a carbonate-enriched horizon, not a clay-enriched horizon. The topography on which this soil occurred was level and the vegetation was a wormwood—meadow grass (*Poa*) community.

A typical Serozem (Calciorthid) from near Isfahan in Iran (Dewan and Famouri, 1964) had a 1 cm thick desert pavement of fairly angular volcanic rocks overlying a loose, light brownish-gray, coarse sandy loam 4 cm thick. This was underlain, in turn, by a brown loamy coarse sand, with abundant gypsum crystals appearing below 40 cm. The calcium carbonate content was 25% in the 4 cm thick surface horizon, 24% in the next horizon, and 14% in the gypsum horizon below 40 cm. Salinity was low in the upper horizons, but moderate in the gypsum layer. Dewan and Famouri note that Iranian Serozems are calcareous throughout the entire profile, with some lime accumulation near the surface. The organic-matter content is almost always less than 0.5%, the base saturation of the cation-exchange complex is high, and the pH of the saturated paste is about 8.

Vegetation, where it is present, consists of a thin cover of desert shrubs and grasses. *Alhagi camelorum* is the dominant shrub at the lower elevations and *Artemisia* is the main shrub at higher elevations. Land surfaces are level.

Iranian Serozems resemble Calciorthids; they do not appear to have mollic epipedons or argillic horizons in the profiles.

No good information is available on the soils of the Arabian peninsula. Dudal (1963) has prepared a generalized map of the Near East but there is no descriptive material accompanying the map. It is likely that Lithic Torriorthents (shallow undeveloped soils over solid rock) are commoner than Calciorthids on sloping land because erosion is greater and development of a calcic horizon would be very slow. Paleorthids (soils having a petrocalcic horizon within 1 m of the surface) are widespread in the Eastern Province of Saudi Arabia (Chapman, 1971) and probably are extensive throughout the peninsula on level piedmont slopes. As would be expected in low-rainfall areas, calcareous soils dominate the landscapes.

Calciorthids in Syria, Iraq, Afghanistan, Pakistan, and India are similar to those in Iran. Gray is the dominant color, soils are coarse textured and shallow to moderately deep, in the main, and surfaces are covered with a desert pavement. Torriorthents are Orthents that show little or no evidence of development of pedogenic horizons, occur primarily on recent erosional surfaces, have a torric (dry) moisture regime or are saline in the upper part of the solum, and have a soil temperature regime warmer than cryic. Soil reaction (pH) is usually neutral, soils are commonly calcareous, the topography is moderately to steeply sloping and the vegetation consists of sparse shrubs and ephemeral grasses and forbs.

In Asia, Torriorthents which are shallow in depth over rock (Lithic Torriorthents) are found mostly on the moderate to steep slopes of barren desert mountains.

Dewan and Famouri (1964) described Lithic Torriorthents on 15—40% slopes in Iran as consisting of, at 0—15 cm, a brown, slightly gravelly strongly calcareous loam. At 15—40 cm, a very pale-brown, very strongly calcareous loam was present. Below 40 cm, the parent material was calcareous conglomerate rock. Limestone outcrops are common in these Lithic Torriorthent areas. Torriorthents grade into Calciorthids and Camborthids.

Torrifluvents are Fluvents that have a torric moisture regime and are not flooded frequently or for long periods. Most of them are alkaline, many are calcareous, some are saline, and they usually occupy level areas on river flood plains. The vegetation consists mainly of xerophytic shrubs but many Torrifluvents have more or less dense stands of *Tamarix* (salt cedar). Soil textures vary from gravelly to clay, and the soils frequently are stratified, with strata differing widely in texture. In the wadis, coarse textures dominate; on the permanent streams, the large deltas tend to be fine textured whereas coarse textures are prevalent in the upper reaches and in the fans of side drainages.

In this association, Paleorthids (Calciorthids with a petrocalcic horizon) are of significant extent in the Arabian Desert. Torripsamments (sand dunes) can be found throughout the association and Salorthids (saline depressions) also are present.

AR2. Camborthids — Torriorthents — Gypsiorthids

This association is located in three plateau areas in Soviet Central Asia: the Ust Urt Plateau between the Aral and Caspian seas, a plateau on the east side of the Aral Sea, and several small plateaus in the Karshi region of southwestern Kyzyl Kum. The underlying material is mainly limestone, sandstone, shale, and marl on the Ust Urt Plateau and the Karshi plateaus, and sandstone and gravelly sands on the Aral plateau. Beds of columnar gypsum are common in the subsoils of the Ust Urt and Karshi plateau soils and occur sporadically in those of the Aral plateau. While the overall aspect of the plateaus is planar, there are ridges, depressions, and sand dunes spotted around the landscape. The flat depressions are fine textured and contain Haplargids and saline Torriorthents. Ridges frequently are gravelly or stony. The vegetation consists of xerophytic shrubs and grasses.

Camborthids are Orthids having only a cambic horizon. In a cambic horizon, soil-forming processes have changed the original mineral matter enough to produce evidence of structural development — if the soil contains adequate clay — but not enough to produce any other diagnostic horizons. The weak development exemplified by cambic horizons can result in the formation of free iron oxides and a small amount of silicate clays, as well as in the redistribution of carbonates and the obliteration of most of the original mineral structure. The cambic horizon may be at the surface, if erosion has occurred, but commonly is immediately below one of the diagnostic epipedons. It is part of the solum and lies in the position of a B horizon. In some classification systems, it is called a B horizon. Camborthids are in between, from the standpoint of profile development, the Entisols and other Orthids and Argids. Soils coarser in texture than loamy fine sands cannot be Camborthids.

Camborthids on the plateaus in the association frequently have gypsum in their profiles. Columnar gypsum underlies many of them within 1 m of the soil surface, especially on the Ust Urt and Karshi plateaus. Lobova (1960) says that the columnar gypsum is a relict of geologic processes operative in the Pleistocene rather than being the result of present-day soil-forming processes. Gypsum also occurs as veinlets in old root channels, as segregations in small spots, as coatings on the under side of gravels and stones, and as a salt disseminated throughout the soil mass. Camborthids grade into Torriorthents and Calciorthids.

A typical Camborthid on the Ust Urt or Karshi plateaus has a gravelly desert pavement underlain by a loam or sandy loam, with or without gravel, to a depth of 90 cm or more. Carbonate "white eyes" (spots) are prominent between about 15 and 50 cm, coinciding with the cambic horizon. Carbonates generally are highest in the surface soil and decrease in concentration with depth. Gypsum segregations and columns begin somewhere below 40 cm but the profile may have a small amount of gypsum above that point. Camborthids on the plateau east of the Aral Sea are similar to those on the other plateaus except that gypsum is less likely to be found within 1 m of the soil surface. In all the Camborthids, salinity increases with depth, beginning in the upper subsoil.

Torriorthents in this map unit are of two kinds. Those occurring in the many flat depressions are deep, fine textured, and saline. The soils are undeveloped or very weakly developed and the salinity is derived from the shale parent material. There is no water table within several meters of the surface. Takyrs (Haplargids) are found in some of the depressions. A second kind of Torriorthent has formed on gravelly sandy slopes and over limestone and sandstone. Those on gravelly slopes are moderately deep to deep while those on limestone and sandstone are shallow. All of them are more or less calcareous and many are gypsiferous.

Gypsiorthids in this association have a calcareous coarse to medium textured surface soil overlying beds of columnar gypsum at 15—40 cm. Carbonate content frequently decreases with depth in the gypsum beds.

Torripsamments, Haplargids, Calciorthids, and Natrargids are other great groups occurring in this association but, while common, they are not extensive.

AR3. Salorthids — Torriorthents

Salt flats of inland depressions and coastal plains are the locale for this group of soils. Runoff water flows into the depressions and creates wet conditions for at least part of the year. The water table is close enough to the surface during much of the year for capillary movement to bring water and dissolved salts to the surface. Salt crusts of variable thickness are common. The salt in the depressions comes primarily from solution of the underlying evaporites and of exposed salt domes, whereas that of the coastal plains arises from saline groundwater, deposits of cyclic salts or, a combination of both sources. While the overall aspect is planar, there are low and high places and intermingled wet and dry surfaces. Vegetation is sparse in the salt crust areas, but stands of *Tamarix* and other salt-tolerant shrubs, forbs, and grasses can be found on the less-saline soils with a high water table.

In the Dasht-i-Kavir, the largest salt desert in the world, there are thick salt crusts strong enough to support vehicle travel over the surface, but there also are places where the crust overlies wet soil and is unable to support human traffic. When the salt crust forms hard, sharp ridges, passage over them may be impossible by vehicles and very difficult for humans (Krinsley, 1970). In playas where frequent flooding occurs, the surface may be free of salt and the familiar polygonal cracks characteristic of sodium-affected soils appear when the surface dries. Conditions similar to those in the Dasht-i-Kavir occur in the salt marsh on the Iran—Afghanistan border into which the Helmand and other rivers flow.

Salt domes are a unique feature of Iranian deserts. Inextensive in area, they are important sources of salt for the plains surrounding them due to the salt carried in runoff water from the domes. Dewan and Famouri (1964) estimate that the domes cover 1% or less of the land surface of Iran, but are responsible for salinity problems in about 7% of the nation.

Krinsley (1970) noted the presence of white salts and black salts on the surface of soils of the Dasht-i-Kavir. The black salts are much less common than the white salts. Although no chemical analyses of the salts are given, it seems likely that the black salts are deliquescent calcium and magnesium chlorides, whereas the white salts probably are the usual mixture of sodium, calcium, and magnesium chlorides and sulfates. Black salts are found in other countries of Asia, notably near the Arabian Sea coast in Pakistan, the Baku region on the Caspian Sea, and in the Lower Mesopotamian Plain of Iraq, as well as in Africa and North and South America.

Saline soils of the low-lying coastal regions around the Caspian Sea do not generally exhibit the thick salt crusts which are typical of the Iranian salt flats. In these soils, salts tend to be distributed more evenly throughout the profile and any crusts which do appear are relatively thin.

Dewan and Famouri (1964) describe a Salorthid in the Dasht-i-Kavir as having the following characteristics:

Garmsar silty clay loam, level or nearly level (0—1%)

0—20 cm very dark gray-brown (10YR 3/2 m) heavy silty clay loam, with friable consistence and granular structure, high in organic matter, abundant mottling, abundant roots, and severe salinity

20—60 cm very dark gray to very dark gray-brown (10YR 3/1.5 m) clay loam, massive structure, sticky when wet, abundant mottling, and high in roots, medium salinity

60—100 cm dark olive-gray (5 Y 3/2 m) clay, massive structure, very sticky when wet, abundant mottling, and some roots, medium salinity

100—200 cm dark gray-brown (2.5 Y 4/2 m) heavy clay, plastic when wet, abundant mottling, some roots, medium salinity

200—300 cm dark olive-gray (5 Y 3/2 m) clay, plastic when wet, abundant mottling, some roots, medium salinity

The abundant mottling (orange and red spots against a darker background color) in the profile attests to the prolonged wetness of the soil. Other Salorthids in the better drained parts of the Dasht-i-Kavir do not show so much mottling as this soil, if any, because the water table is close to the surface for shorter periods of time.

Salt marsh Salorthids cover large areas in the depression (Hanun-i-Helmand) receiving the runoff from the southwestern Afghanistan rivers, as well as in the marshy regions of the area delineated along the northeast shore of the Caspian Sea and in the Rann of Kutch in India. Salorthids in the other Caspian Sea areas and in the Indus Delta of Pakistan have water tables at one or more meters below the soil surface most of the time.

The natural vegetation of the saline soils is salt tolerant shrubs, sedges, reeds, and grasses. *Tamarix*, *Carex*, *Phragmites*, *Salicornia*, and *Juncus* are widespread.

Torriorthents of the saline variety occur on the higher and generally more sloping land around salt marshes. The underlying geologic material from which the soils formed is gypsiferous and saline marls. Virtually all, if not all, of the Torriorthents are calcareous throughout the profile, of variable texture but usually coarse, less than 1 m deep over the parent material, and low in organic matter, which decreases in amount with depth.

Natrargids and Torrifluvents are present but are much less extensive than the Salorthids and Torriorthents.

AR4. Salorthids — Gypsiorthids — Calciorthids

Arid coastal plains, ancient sedimentary plains, and recent alluvial plains are the surfaces upon which these soils have developed. The largest single area extends from Syria through Iraq to the coasts of the Persian Gulf and the Arabian Sea. Another long and narrow delineation begins in the Rift Valley of Israel and Jordan, then continues along the coast of the Red Sea and the Gulf of Aden. In the coastal areas, alluvial fans provide topographic relief where mountains line the coast, and sand dunes can be found in many places. Secondary calcium carbonate and gypsum crusts are extensive in Syria and Iraq where inland seas were formed and layers of salts were deposited as the seas dried up. Scattered xerophytic shrubs are the only plants found over large expanses of the dry soils, but the vegetative cover may be heavy in the salt marshes of the Tigris—Euphrates delta. *Suaeda fruticosa* is the dominant shrub on the coastal plains.

Salorthids are both natural and man-made in this association. The naturally occurring ones are located in the Tigris—Euphrates delta, on the coastal plains, and in some interior depressions that are flooded periodically. The man-made ones are extensive in the presently and formerly irrigated lands of the great Tigris—Euphrates alluvial plain of Iraq and Iran called the Lower Mesopotamian Plain (Buringh, 1960). As in many alluvial soils, stratification is a common feature, with textures varying from fine sand to clay. Soils are deep and silty clay loam is the dominant textural class in the Lower Mesopotamian Plain. Soil permeability is low. All of the soils are strongly calcareous and gypsum can be found somewhere in the profile of nearly every soil. The water table may be within 1 m of the surface, as it is in the greater part of the plain, or it may be two or more meters down. Buringh (1960) notes that salinity decreases with depth in the nonirrigated soils but increases with depth under irrigation. The topography is flat, with micro-relief differences of only a few meters, at most, and soils are variable over short distances horizontally. In short, while the general appearance of the physical features of the plain indicates homogeneity, there are large soil and moisture differences from place to place.

A noteworthy condition in Salorthids of the Lower Mesopotamian Plain is the presence of *sabakh* soils (Buringh, 1960). These are saline soils having a high percentage of deliquescent salts (calcium and magnesium chlorides) and magnesium sulfate. Wherever they occur, the surface is always moist and darker colored than the normal saline soils of the surrounding area. In the United States, similar saline soils have been called "black alkali" soils because of their dark color, although they contain little exchangeable sodium (Kelley, 1951). They differ markedly from the usual black alkali (sodic) soils in which the exchangeable sodium percentage is high.

Salorthids of the coastal plains are generally coarse to medium textured in the surface, fine textured in the subsoil, slowly permeable, and highly saline on the seaward side of the plain, becoming coarser textured inland. Salinity is high in the surface horizon and throughout the profile. The water table usually is one or more meters below the surface. Practically all of the soils are deep and calcareous; most of them have gypsum in the subsoil. Sand dunes have formed over many of the saline flats.

Gypsiorthids grade into Calciorthids and Torriorthents over much of the area in this association. In some places, gypsum crusts are on the soil surface, whereas in other places concentrated gypsum horizons are at varying depths below the surface. Petrogypsic horizons are common in the gypsum desert of Iraq and Syria which lies mostly west and north of the Euphrates in Iraq and north of the river in Syria. Surfaces are coarse textured where the petrogypsic layer is not exposed, and both the soils and the gypsum deposits are strongly calcareous.

Calciorthids are found on the alluvial fans which are extensive between the mountains and the coastal plains of the Red Sea and the Iranian side of the Persian Gulf. They are coarse textured, have a desert pavement of gravels, are calcareous throughout the profile, and have a calcic horizon within 50 cm of the surface.

Other great groups represented in this association include Torriorthents, Torripsamments and Torrifluvents. There probably are some Camborthids, as well.

AR5. Calciorthids — Salorthids — Haplargids

Stretching across the northern limit of the Aridisols, from the Caspian Sea to eastern Mongolia and north central China, these soils represent the transition between less-developed Aridisols of the south and the Mollisols of the wetter and colder north. They are on sloping or level plains, with numerous depressions, stream channels, some desert mountains and low hills, and scattered small sand dunes. Upland soils of the plains are Calciorthids; soils of the depressions are Haplargids or Salorthids. Vegetative cover is better on the north side of this soil zone than on the south side. Several species of *Stipa* and *Agropyron* provide a fairly heavy grass cover in the north, with *Artemisia* the principal shrub. To the south, grass thins out and shrubs such as *Salsola* and *Artemisia* increase in number. *Haloxylon* is common on sandy soils and sand dunes. Depressions are either barren of vegetation or have salt-tolerant species growing in them.

Calciorthids are similar to those of Association AR1 except that they customarily have a higher organic-matter content, the calcium carbonate content is lower, gypsum usually is absent from at least the upper 2 m of soil, and the vegetative cover is greater and contains more grasses. In general, the soils are grayish brown, moderately deep to deep, coarse to medium textured, frequently gravelly, calcareous to the surface, have a calcic horizon at 25—75 cm (sometimes two carbonate horizons), are nonsaline, and have a desert pavement (Bespalov, 1951).

Salorthids are extensive in the depressions of Mongolia and China where the water table is close to the surface for much of the year. The source of the salts is primarily the saline geologic material underlying the recent surficial deposits in the depressions. Other sources are salt carried in runoff water from the surrounding uplands and, possibly, salts carried in the clouds of dust which obscure the atmosphere periodically. Bespalov (1951) mentions salt efflorescences of 1—3 cm in thickness being removed in a single wind storm. Deposition of the salt downwind could easily have a significant effect on the salinity of the soils after many years of accumulation.

Most of the Salorthids are deep, medium to fine textured, calcareous to the surface, contain little gypsum and are mottled. They usually are barren of vegetation or have a thin growth of *Salsola*.

Haplargids in this association are the takyrs of Central Asia. Takyrs are soils of closed basins and are notable for the almost complete absence of higher plants and the presence of an algal and lichen crust. Higher plants are absent because takyrs are so slowly permeable that runoff water from the surrounding uplands stands on the surface for long periods of time, drowning any plants that might become established. Between rains, the water evaporates, leaving a hard dry crust with polygonal cracks 2—10 cm deep. Lobova (1960) describes a typical takyr as having the polygonal crust, a pinkish or pale yellow-gray surface color, and, frequently, a cover of desert papyrus (thin sheets of algal crusts, rolled up on the edges). Immediately below the surface, the first 3 cm is a coarsely porous (honeycomb) loam to clay loam horizon. The next 3—5 cm is a laminated noncompact horizon of the same texture. Beneath that is the beginning of a more compact and clayey moderately developed argillic horizon. Illite, chlorite, and montmorillonite are the dominant clay minerals.

Takyrs usually have a nonsaline crust, the underlying soil being more or less saline to a considerable depth. There is no water table within several meters of the surface. Gypsum occurs throughout the profile, with a maximum under the crust and a second layer of gypsum accumulation in the lower part of the argillic horizon. Exchangeable sodium is highest in the surface and generally decreases steadily with depth. The entire profile is calcareous. Soil pH is strongly alkaline in the surface and slightly alkaline to neutral in the subsoil.

Among the variations in the properties of takyrs is the frequent absence of a clearly defined argillic horizon, sandy textures rather than loamy or clayey, high amounts of salts and exchangeable sodium in the subsoils, and low levels of soluble salts in the upper 20—50 cm.

Many takyrs probably belong to the Camborthid rather than to the Haplargid great group since apparently it is not unusual for the argillic horizon to be absent. Some of the takyrs may be Natrargids.

Additional great groups include Torriorthents, Torripsamments, and Paleorthids.

Entisols

Psamments, Orthents, and Fluvents are the main suborders among the Entisols of Asia. Psamments (sand dunes) are of great extent in four distinct areas: the Rub' al Khali in Saudi Arabia, the Kara Kum and Kyzyl Kum in the U.S.S.R., the Thar (Rajasthan) Desert in India, and the Takla Makan Desert in China. Orthents are the shallow soils of the desert mountain and saline rocks and the undeveloped soils of alluvial fans. Fluvents are the undeveloped soils of flood plains and deltas of intermittent and permanent streams.

E1. Psamments — Haplargids — Natrargids

Vast expanses of towering sand dunes dominate the landscape of this association, some of them over 100 m in height, many of them completely barren of vegetation. Interdune areas can consist of a few feet of sand overlying various kinds of rocks, sedimentary deposits, and old soil surfaces or they can be free of sand. In the latter case, Haplargids and Natrargids frequently are the soils occurring between the dunes because clayey deposits seem to underlie many dune fields. There are no extensive areas of level sandy soils in Asia; where sand occurs, it usually piles up in large or small dunes. Perhaps as much as half of the Asian sand dunes are of the unstable, shifting type, with little or no vegetative cover. Where vegetation is present, it usually is sparse and consists mainly of *Haloxylon*, *Calligonum*, *Ammodendron*, *Carex* and *Bromus*. Sand-dune ridges are more likely to be bare of vegetation than are the sides and the interdune spaces.

Psamments are undeveloped sandy soils. They may be noncalcareous on the surface but, if so, generally are calcareous throughout the remainder of the profile. Some segregated lime may be found one or more meters below the soil surface but gypsum rarely is present unless the dunes were formed by deflation of gypsiferous materials in nearby depressions or gypsum beds. Textures are loamy sand or coarser, the soils are almost always nonsaline, and structural development is generally nonexistent, leaving the soil as a loose mass of individual particles subject to easy movement by wind.

Haplargids are Aridisols with a thin argillic horizon, whereas Natrargids have a natric horizon. The two great groups occupy the same physiographic position: interdune areas. Most of them formed on ancient alluvial plains which now have been encroached upon by sand dunes. Haplargids and Natrargids of the sand dune areas are medium to fine textured, deep, calcareous, and slightly to moderately saline. The Haplargids form when the soils are permeable and there never has been a high water table; Natrargids are less permeable and probably were quite saline at one time, either because the alluvium was saline or because of a high water table. The water table is deep under both Haplargids and Natrargids.

Torriorthents, Salorthids, and Torrifluvents are present in this association but they are not widespread.

E2. Torrifluvents — Salorthids

The five delineated areas in this association are alluvial flood plains of permanent streams: the Indus River in Pakistan, the Tarim and Cherchen rivers in Sinkiang, the Tedzhen and Murgab rivers in the Turkmen S.S.R., and the Kura and Araks rivers in the Azerbaijan S.S.R. Irrigation is practiced or has been practiced in all of these areas; the least development has occurred in the Tarim and Cherchen river valleys. The oasis of Mary in the Murgab River basin was a major stopping point on the ancient caravan route between China and southwest Asia. High water tables at some time of the year is a problem which has arisen due to seasonal flooding of the rivers but also to the application of excess amounts of irrigation water. Where the water table is high for several months of the year, Salorthids appear. In the well-drained part of the flood plain, Torrifluvents are dominant. The vegetation is cultivated crops, by and large, with *Tamarix* and *Populus* along canals and rivers and salt-tolerant grasses, sedges, and forbs in fields which were abandoned when the soils became excessively saline.

Torrifluvents are the young, undeveloped, usually stratified soils of the flood plains of rivers. Salorthids are the same soils but with a water table close enough to the surface for capillary movement of the water to bring salts to the top. Soil textures are highly variable within the profile and from place to place within each basin. Silt loams and silty clay loams probably are the commonest surface-soil textures, especially at some distance from the river channels and at the lower ends of the basins.

When the soils are well-drained Torrifluvents, crop production generally is good on the finer-textured soils and poorer on the sandy soils. On the Salorthids, crop growth is adversely affected to varying degrees, depending upon how much salt has accumulated in the upper part of the profile, the salt tolerance of the crops, and soil and water management practices. Large tracts of land in the Murgab and Tedzhen river basins have been abandoned because the salinity level became intolerably high; salinity is credited with forcing the abandonment of as much as 20,000 ha of irrigated land each year in the Indus plain.

In at least three of the areas, deliquescent calcium and magnesium chlorides are present in large enough quantities to give the typical dark appearance to affected soils. The three places where this kind of "black alkali" has formed are the Indus plain, the Murgab basin, and the Kura-Araks plain southwest of Baku. A high water table seems to be essential — worldwide — for the accumulation of calcium and magnesium chlorides in preference to sulfate salts and sodium chloride, but a high water table does not mean that such salts will necessarily be the dominant ones.

Other great groups occurring in this association are Ustifluvents, Torripsamments, and Torriorthents.

E3. Cryorthents — Salorthids

Little can be said about the Tibetan Plateau where this association is mapped because practically nothing is known about its soils. There is some doubt whether Cryorthents actually are present, but there is little doubt that the widespread occurrence of Salorthids in depressions. Ma and Wen (1958), in their generalized soil map of China, call the soils of Tibet Alpine cold-desert soils but they do not provide any information on the soils or the environmental conditions other than to say that the elevations generally exceed 3,000 m. Kovda and Lobova (1971) call the soils high-mountain desertic soils and high-mountain cryogenic-meadow saline soils.

Presumably, then, the *Cryorthents* would be undeveloped soils of mountains, alluvial fans and piedmont slopes lying above the intermountain basins where salt lakes and saline soils would be found. They would resemble hot-desert soils in most respects except that the organic-matter content may be higher and the subsoils may contain ice crystals during at least part of the year. The *Salorthids* would be similar to Salorthids in the hot regions although the kinds of salts may be different due to the effect of temperature on solubility characteristics.

E4. Torrifluvents — Calciorthids — Salorthids

This association encompasses the Indogangetic plain of Pakistan and India, a plain filled with alluvium from the rivers flowing out of the Himalayas to the north and having an admixture of sandy eolian material coming from the Rajasthan Desert area to the southwest. The continuation of the Gangetic plain to the east has a humid to perhumid climate and is excluded from this discussion.

Torrifluvents are the dry, undeveloped soils of flood plains. In the Indogangetic plain, they usually are coarse to medium textured, deep, stratified, calcareous, slightly saline, permeable soils. The water table is two or more meters below the soil surface most of the time. When the water table is close to the surface, due to the presence of fine-textured strata or to position of the soil in depressions, Salorthids frequently develop (Raychaudhari and Govinda Rajan, 1972). Torrifluvents are the principal agricultural soils in the Indogangetic plain.

Calciorthids are found on older alluvium where translocation and accumulation of calcium carbonate has occurred. These soils are concentrated on the southern side of the Indogangetic plain and probably represent soils to which calcium carbonate has been added by deposition of wind-transported calcareous particles from the Rajasthan Desert. In other respects, they are similar to Torrifluvents but with less-obvious stratification.

Salorthids in the Indogangetic plain largely are man-made as the result of the application of excess amounts of irrigation water on a plain which is nearly flat and where groundwater drainage is slow. Salts accumulate from the long-continued evaporation of groundwater at the soil surface; the salinity of the river water is low but groundwaters frequently are quite saline. Salorthids are increasing in area as the salinity and drainage problem worsens under the pressure of expanding irrigation.

In addition to the three main great groups described here, there are some Ustifluvents, Haplaquents, Paleustalfs, Haplargids, and Natrargids, according to Raychaudhari and Govinda Rajan (1972).

E5. Torriorthents — Rock Outcrops — Calciorthids

Shallow or gravelly soils on moderately to steeply sloping land typify this association. The landscape is one of barren desert mountains, with narrow valleys in the more rugged country and fairly broad alluvial fans and plains in the less rugged areas. Torriorthents and Rock Outcrops occur on the mountain slopes and the upper part of the gravelly alluvial fans, Calciorthids on the lower part of the fans and on the plains. There probably are Calciustolls (Mollisols with a calcic horizon) and Argiustolls (Mollisols with an argillic horizon) on the older and wetter surfaces bordering the higher mountains, particularly in Iran, Afghanistan, and China.

Torriorthents are the shallow soils over bedrock or deep gravelly soils on slopes. They are usually calcareous throughout the profile. *Rock Outcrops* is a land type rather than a soil because it consists of exposed rocks with no significant soil cover. *Calciorthids* are similar to the deep gravelly Torriorthents but have a calcic horizon within 50 cm of the surface.

Haplargids, Paleargids, Paleorthids, and Torrifluvents are among the other great groups occurring in this association.

Mollisols

The main Asiatic Mollisol zone forms a flattened semicircle around the north side of the Aridisols of the U.S.S.R., Mongolia, and China. A small area in central Turkey is also occupied by Mollisols.

Mollisols are the soils of the grasslands, extensive in the northern hemisphere but inextensive in the southern hemisphere. Agriculturally, these are the soils on which dry farming is practiced if the growing season is long enough and the soils are not too cold. Droughts occur frequently enough in the drier part of the Mollisol zone to be a hazard to crop production. In Asia and Africa, the grasslands are commonly referred to as steppes and in some classifications of climates the arid regions are said to be those drier than the steppes, which is another way of saying that the regions are not wet enough for dry farming to be successful.

The topography of the grasslands varies from level to gently sloping to moderately undulating, with alluvial valleys and some large depressions present. The vegetation consists of perennial grasses such as *Stipa*, *Agropyron*, *Poa*, and *Bromus* and of shrubs, principally *Artemisia* and *Caragana.*

M1. Argixerolls — Haploxerolls — Ustorthents

Xerolls are the Mollisols of Mediterranean climates, with wet winters and dry summers. In Asia, they occur in the Anatolian Plateau of central Turkey. Characteristically, the *Argixerolls* have hard, massive, medium-textured, mollic epipedons; an argillic horizon in or below the mollic epipedon; a high base saturation; and are noncalcareous in the surface but may have carbonate accumulations in the subsoil. Organic-matter content meets the criteria for a mollic epipedon but is relatively low (1—3%). Winter rainfall, although low, is adequate to leach the soils of soluble salts and most carbonates because it falls when evapotranspiration is minimal.

Haploxerolls have either a cambic horizon below the mollic epipedon or have no diagnostic subsurface horizons of any kind. Profiles are similar to those of Argixerolls except for the absence of an argillic horizon. The soils are younger than the Argixerolls and usually are on slopes where erosion has prevented formation of an argillic horizon.

Ustorthents are shallow soils over generally unconsolidated rock. Base saturation is high but carbonates usually are absent in the upper layers. Medium textures are dominant.

Argixerolls are found on level, older surfaces; Haploxerolls on gentle to moderate slopes; and Ustorthents on the steep slopes. Clays are mostly of the kaolinitic type. Internal drainage is good.

Torrifluvents represent the principal other great group.

M2. Argiustolls — Haplustolls — Haploborolls

This association forms a band on the east and north side of the Gobi Desert of Mongolia and China. The climate is wetter and cooler than in the desert and the vegetation is a fair to good cover of grass. Grazing livestock is the principal use of the land; some dry farming and irrigation is carried on in the lowlands.

Argiustolls are the Ustolls with an argillic horizon. They occur on the level plains and terraces between mountain ranges. Organic-matter content is moderate (3—4%) in the mollic epipedon, the surfaces are medium textured and the B horizon fine textured, and the soils are deep and calcareous. A calcic horizon sometimes can be found beneath the argillic horizon.

Haplustolls in this area generally have a cambic horizon as the result of the beginning of clay illuviation and structural development in the subsoil. Medium textures are prevalent and the soils are moderately deep to deep and calcareous. They occupy the gentle to moderate slopes of the low ridges on plain surfaces and of the footslopes of mountains.

Haploborolls are the cold Mollisols. The organic-matter content is relatively high (5% or more), subsoils are moist for several months of most years, the surfaces are noncalcareous, and the base saturation usually is relatively low for Mollisols. They occupy depressions in the undulating and hilly parts of the landscape.

Ustifluvents, Udifluvents and Natrustolls are other great groups found among the soils of this association.

M3. Haploborolls — Argiborolls

Adjoining the previous association is one with similar soils but a colder soil temperature regime. The grass cover is denser than on the Ustolls and dry farming is more successful year after year, with fewer droughts. Borolls usually are moist for a greater period of time during the year than are the Ustolls.

Haploborolls have a moderately high organic-matter content, profiles usually are medium to fine textured and have a cambic horizon, the soils sometimes are calcareous in subsoil, and the base saturation is moderate. The typical ones occupy the gently sloping surfaces and low ridge tops, but those occurring in depressions have higher organic-matter contents and a thicker mollic epipedon.

Argiborolls have an argillic horizon which restricts water movement and sometimes results in drainage problems, particularly since they occur on rather level surfaces. They have moderate base saturation and are non-calcareous in the mollic epipedon. Surface soils are medium to fine textured.

M4. Argiustolls — Natrustolls

North of the Aridisol soils in Soviet Central Asia is an extensive plain in which the climate becomes progressively colder and wetter toward the north. Similar soils are found in the M4 and M5 areas; the difference is largely in the soil temperature regime, with Ustolls having a mesic or warmer regime and Borolls a frigid or cryic soil temperature regime. Argiustolls occupy the higher parts of the plains and Natrustolls the lower slopes and depressions. Natrustolls and Natriborolls are the Russian Solonetz soils.

Argiustolls are the argillic Ustolls without a natric horizon. As in Mongolia and China, the Argiustolls have a moderate organic-matter content, medium-textured surface soils, fine-textured argillic horizon, are deep and calcareous, and may have a calcic horizon beneath the argillic horizon. The soils are non-saline.

Natrustolls are Ustolls with natric horizons. They possess the major characteristics of solonetz soils: a fine-textured, slowly permeable, columnar or prismatic B horizon containing more than 15% exchangeable sodium or, in some cases, less sodium but more exchangeable magnesium. Tops of the columns in the B horizon commonly are rounded, the bottoms are flat, the color is dark, and the faces of the columns are shiny (Joffe, 1949). Gypsum veins frequently are found in the lower part of the B horizon. Surface soils generally are noncalcareous but subsoils are calcareous.

Ustifluvents and Udifluvents are the principal other great groups.

M5. Argiborolls — Natriborolls

This association lies above the M4 association and reflects the lower temperatures occurring there. The grass cover is denser, the growing season is shorter, and droughts are somewhat less frequent, making dry farming less hazardous insofar as moisture supply is concerned.

Argiborolls and *Natriborolls* are similar to the Argiustolls and Natrustolls of the adjoining association. Organic-matter content of the Borolls usually is a little higher than that of the Ustolls. The Natriborolls are Solonetz soils and have the typical columnar subsoils with rounded tops, flat bottoms, dark color, and shiny faces on the columns of the natric horizon. Natriborolls are less permeable than the Argiborolls, but the latter are only moderately permeable.

Vertisols

V. Pellusterts — Chromusterts — Ustorthents

Vertisols of different types cover the extensive Deccan and Malwa plateaus in India and the lowland plains bordering the plateaus. The principal underlying — and frequently exposed — rock constituting the two plateaus is basalt. The Pellusterts and Chromusterts are called black cotton soils or regur soils in India. The plateaus consist of rocky hills, valleys, undulating slopes, and smooth plains. Below them, the topography is undulating to smooth. The climate is semiarid to subhumid, with a high annual rainfall of 750—1,000 mm concentrated in July and August and preceded by little or no rain during the long dry season before the monsoon begins.

Pellusterts and *Chromusterts* have similar profiles, the difference between them being in the color of the upper 30 cm: the darker ones are Pellusterts, the lighter ones Chromusterts. The principal identifying characteristic of Vertisols is the deep cracking which occurs during the dry season. Cracks must be 1 cm or more wide at a depth of 50 cm during part of most years for a soil to qualify as a Vertisol. The other limiting property is the presence of at least 30% clay in all horizons down to 50 or more cm. For cracking to appear, the clay must be of the strongly expanding and contracting type, which is why these soils are high in montmorillonitic clays. In many Vertisols, the pronounced expansion and contraction of the clay forms a *gilgai* relief of recurring small (in height) mounds and depressions.

Srinivasan et al. (1969) describe a Chromustert as having a very dark grayish-brown clay throughout the profile, slightly calcareous from the surface down to the parent material, a granular surface, and a blocky subsoil. Below 150 cm, unconsolidated basalt occurs. Some Usterts are noncalcareous on the surface and calcareous in the subsoil. There is no evidence of clay movement in the profile, the organic-matter content is low despite the dark color, and the dominant exchangeable cations are calcium and magnesium. In India, Chromusterts tend to be found mostly in the north of the Vertisol zone and Pellusterts in the south (Srinivasan et al., 1969).

Vertisolic soils in India vary greatly in depth to the underlying unconsolidated rock. The shallow ones (less than 25 cm deep) are *Ustorthents* because they are not deep enough to qualify as Vertisols even though their texture is clay and they have the properties of Vertisols except for the depth of the cracks.

Within the Vertisolic region are areas where red and black soils are intimately mixed. The red soils are Alfisols derived from granite and granitic gneisses (Raychaudhari and Govinda Rajan, 1972) and range from coarse to fine in texture. They differ from the black Vertisols in being higher in potash minerals, lower in clay content, lower in cation-exchange capacity per gram of clay and higher in kaolinitic-clay minerals. Presumably, the red soils are more permeable than the black soils.

Pelluderts and Ustochrepts are also present in this association.

REFERENCES

Bespalov, N.D., 1951. *Pochvy Mongol'skoi Narodni Respubliki.* (*Soils of Outer Mongolia.* Issued in translation by the Israel Program for Scientific Translation, Jerusalem, 1964, 320 pp.; also cited as OTS 64-11073.)

Buringh, P., 1960. *Soils and Soil Conditions in Iraq.* Ministry of Agriculture, Republic of Iraq, Baghdad, 322 pp.

Chapman, R.W., 1971. Climatic changes and the evolution of landforms in the Eastern Province of Saudi Arabia. *Geol. Soc. Am., Bull.*, 82: 2713–2728.

Dewan, M.L. and Famouri, J., 1964. *The Soils of Iran.* FAO, Rome, 319 pp.

Dudal, R., 1963. *Soil map of the Near East.* FAO, Rome, scale: 1:5,000,000.

Joffe, J.S., 1949. *Pedology.* Pedology Publications, New Brunswick, N.J., 662 pp.

Kelley, W.P., 1951. *Alkali Soils.* Reinhold, New York, N.Y., 176 pp.

Kovda, V.A. and Lobova, E.V. (Editors), 1971. *Soil map of Asia.* USSR Academy of Sciences, V.I. Lenin All-Union Academy of Agricultural Sciences, Moscow, scale 1:6,000,000.

Krinsley, D.B., 1970. *A Geomorphological and Paleoclimatological Study of the Playas of Iran.* U.S. Geological Survey, Washington, D.C., part I: 329 pp. (text) and part II: 330—486 (maps and other illustrative material).
Lobova, E.V., 1960. *(Pochvy Pustynnoi Zony SSSR.* (*Soils of the Desert Zone of the U.S.S.R.* Issued in translation by the Israel Program for Scientific Translations, Jerusalem, 1967, 405 pp.; also cited as TT67-51279.)
Ma, Yung-tsu and Wen, Chen-wang, 1958. The principle of soil classification. Basis of the agricultural development. *Acta Pedol. Sin.*, 6: 157—176.
McGinnies, W.G., 1968. Appraisal of research on vegetation of desert environments. In: W.G. McGinnies, B.J. Goldman and P. Paylore (Editors), *Deserts of the World.* University of Arizona Press, Tucson, Ariz., pp.381—566.
Meigs, P., 1953. World distribution of arid and semi-arid homoclimates. In: UNESCO, *Reviews of Research on Arid Zone Hydrology. Arid Zone Res.*, I: 203—210.
Raychaudhari, S.P. and Govinda Rajan, S.V., 1972. Soil genesis and soil classification. In: J.S. Kanwar and S.P. Raychaudhari (Editors), *Review of Soil Research in India.* Indian Council of Agricultural Research, New Delhi, pp.107—135.
Rozanov, A.N., 1951. *Serozemy Srednei Azii.* (*Serozems of Central Asia.* Issued in translation by the Israel Program for Scientific Translations, Jerusalem, 1961, 550 pp.; also cited as OTS 60-21834.)
Srinivasan, T.R., Bali, Y.P. and Tamhane, R.V., 1969. Placement of black soils of India in the Comprehensive Soil Classification System — 7th Approximation. *J. Indian Soc. Soil Sci.*, 17: 323—331.
U.S. Department of Commerce, 1969. *Climates of the World.* Environmental Data Service, Environmental Science Services Administration, U.S. Department of Commerce, Washington, D.C. 28 pp.

Chapter 6

AUSTRALIA

INTRODUCTION

Aridity dominates the land mass of Australia more than that of any other continent. Approximately 75% of Australia is arid according to Meigs' (1953) definition of arid lands (Perry, 1970). The arid region is one large continuous area more than 3,200 km from east to west and 2,000 km from north to south (Fig.6.1). Over most of the continent, the arid section is surrounded by a humid coastal strip in which 95% or more of the population lives, leaving the arid region for pastoral raising of wool, sheep, and beef cattle and — concentrated in the southeast — about 1,000,000 ha of irrigated land.

Aridisols, Entisols, and Vertisols occupy most of the arid lands of Australia (Fig.6.2). Alfisols are significant in the east, south, and west and are locally important soils among the Entisols in the north. There are no large areas of Mollisols. A book on Australian soils prepared for the 9th International Congress of Soil Science in 1968 was the principal source of information on the soils (Stace et al., 1968).

ENVIRONMENT

Climate

Most of the arid zone receives its rain in summer (November to March) but there is a shift to a Mediterranean type of winter rainfall in the southwest. In the north, the arid zone includes areas with as much as 750 mm of rain where a short rainy summer is followed by a long dry winter. Despite the relative wetness of summer in the north, dry periods of several weeks duration can occur between the rains. As is true for other arid regions where precipitation is mainly of the summer type, rainfall is highly erratic and commonly occurs as brief and heavy thunderstorms. The sporadic character of the rains is accentuated in the more arid part of the continent and becomes less prominent in the wetter part.

The lowest annual precipitation is about 100 mm at Mulka in northern South Australia. Table 6.1 presents precipitation and temperature data for representative weather stations across Australia. Alice Springs is in the center of the country with a somewhat continental climate; Bourke is in the north of New South Wales where rainfall is rather evenly distributed throughout

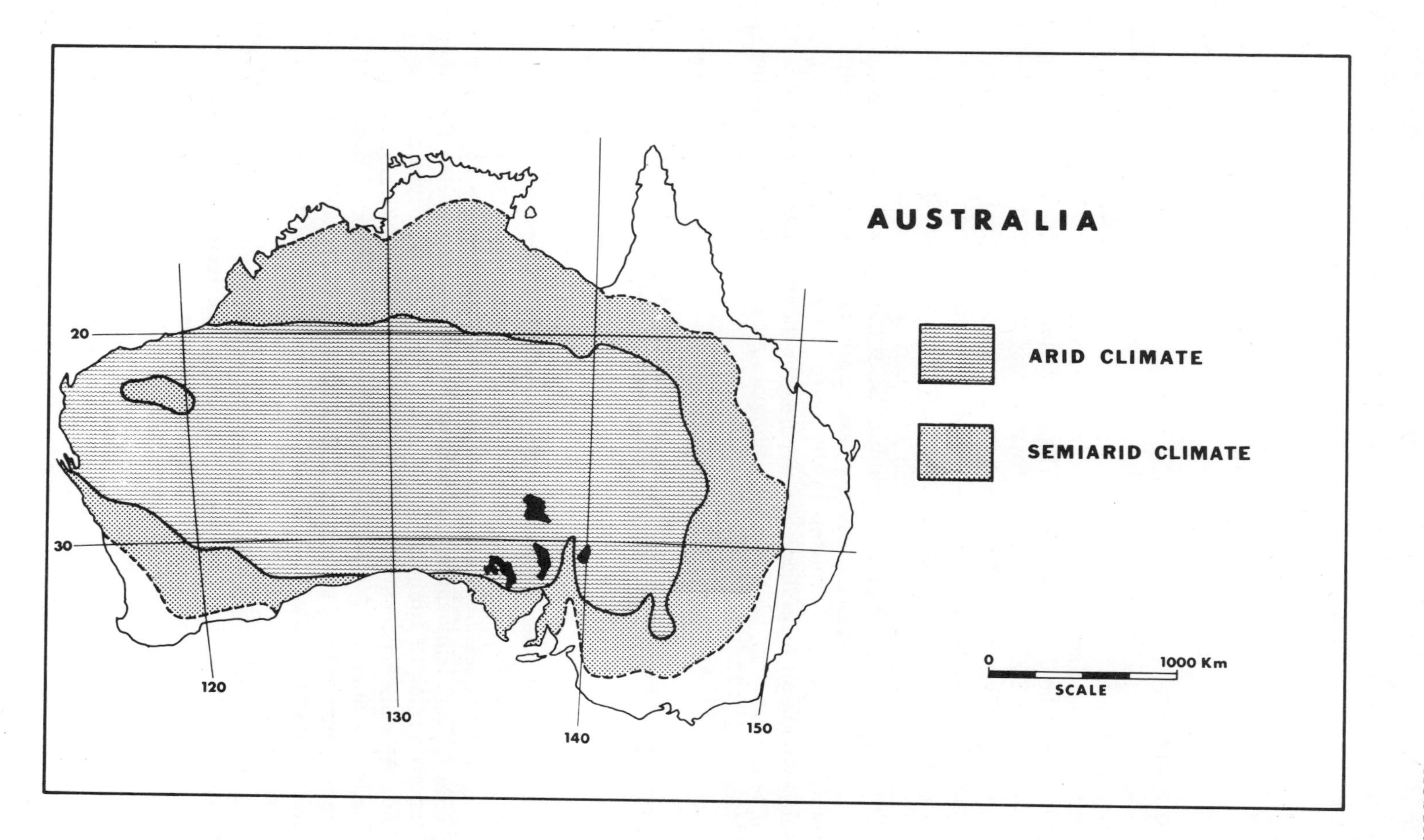

Fig.6.1. Arid regions of Australia (after Meigs, 1953).

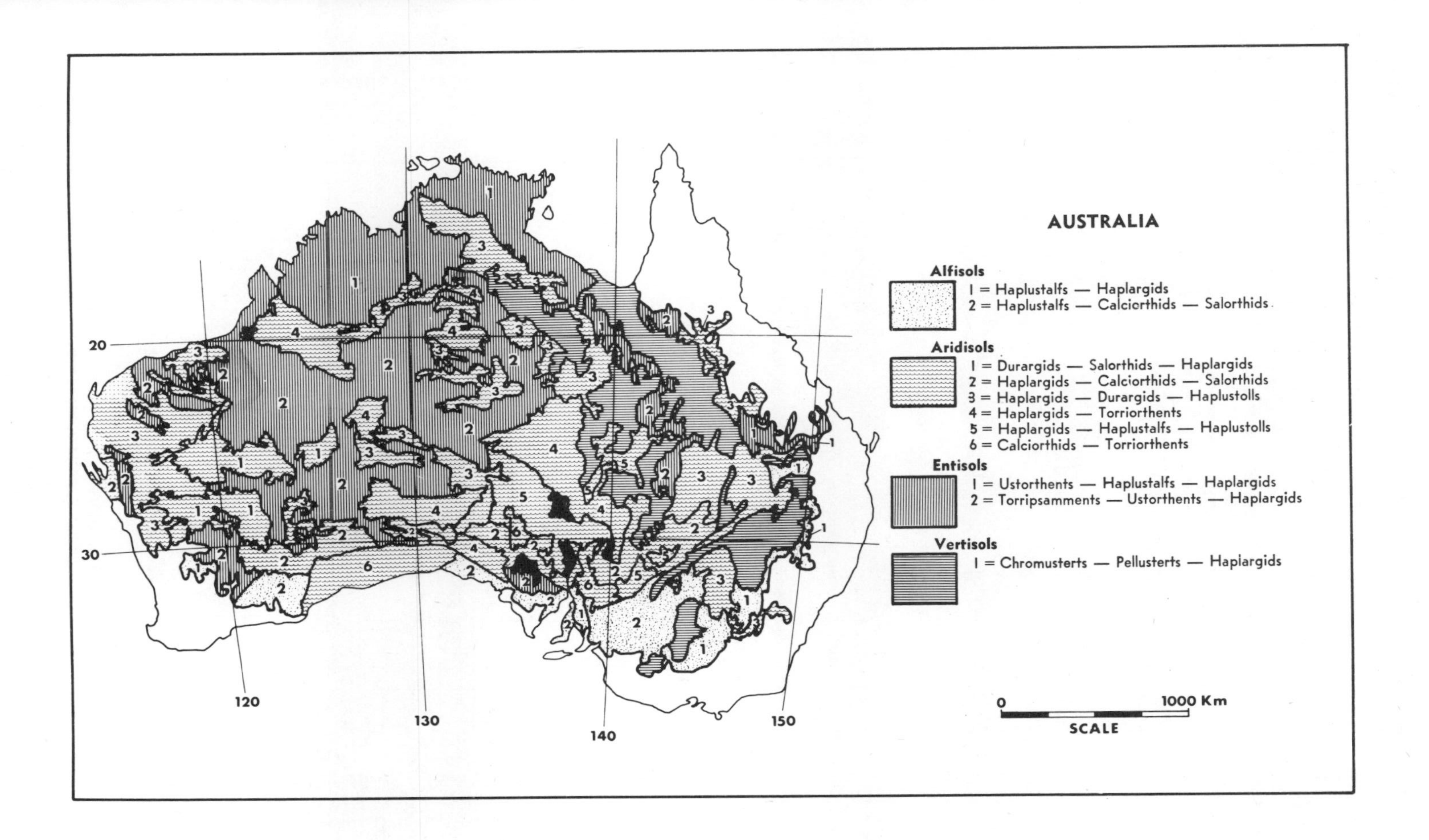

Fig.6.2. Arid-region soils of Australia.

TABLE 6.1

Climatological data for five Australian weather stations

Station	Elevation (m)	Precipitation (mm)													Average annual temperature (°C)
		J	F	M	A	M	J	J	A	S	O	N	D	annual	
Alice Springs	546	43	33	28	10	15	13	8	8	8	18	31	38	253	21
Bourke	110	36	38	28	28	25	28	23	20	20	23	31	36	336	21
Burketown	9	208	160	132	25	5	8	0	0	0	10	38	112	698	26
Carnarvon	4	10	18	18	15	38	61	41	18	5	3	0	5	232	22
Perth	20	8	10	20	43	130	180	170	145	86	56	20	13	881	18

Source: U.S. Department of Commerce (1969).

the year; Burketown lies on the coast of the Gulf of Carpentaria in Queensland and is representative of the wet summer—dry winter hot climate of northern Australia; Carnarvon is located on the west coast of Western Australia; Perth lies outside the arid region in Western Australia.

Temperature ranges are fairly large in every part of the arid region. The record maximum of Bourke is 52°C and the record minimum is −4°C. The least variation in all-time maxima and minima at the stations in Table 6.1 is at Burketown where the respective temperatures are 43°C and 4°C. Frosts are unknown in the north.

Geology and geomorphology

Australia's arid region is an ancient highly weathered land surface which is distinctive among much of the world's arid lands for the age of the soils. According to Mabbutt (1969), the western two-thirds of the arid zone (the Western Platform) has a basement of Archaean granite and gneiss which is exposed in shield areas but is generally covered with erosion products. The eastern one-third is called the Central Basin and consists of marine Mesozoic claystone and soft sandstone, partially covered with Tertiary sands. The Great Artesian Basin of mostly saline groundwater lies in the Central Basin. By about the middle Tertiary Era, 35 or so million years ago, weathering of platform and basin rocks had produced soils with indurated layers called duricrusts. On the platform, the duricrust was made up principally of an iron-oxide laterite; in the basin the duricrust consisted principally of the siliceous material called silcrete. Both crusts are believed to have been formed during a more humid period. Since then, the crusts have been eroded away or now form erosion-resistant caps on tablelands and mesas. When weathered silcrete fragments are deposited on lower slopes, they form a kind of desert pavement called gibber mantles, the term *gibber* referring to rocks.

Laterite and silcrete duricrusts are relicts of a former humid climate; crusts on younger surfaces developed during the later, more arid, climate consist of calcrete (indurated calcium carbonate) or hard gypsum. The latter are typical arid-region indurated layers.

The Australian arid-zone landscape gives a strong impression of flatness; the low and extensive tablelands of the center and west are flat, the mesas of the center and east are flat, the sand plains are flat, the Nullabor calcareous plain is flat, and the large playas and salinas are flat. Only a few mountains such as the Macdonnell Ranges in central Australia provide escape from the monotony of planar features. The highest point in the arid zone is Mt. Zeil, located west of Alice Springs, at 1,500 m above sea level.

Mabbutt (1969) recognizes six desert physiographic types in the arid zone: (1) mountain and piedmont desert (15% of the zone); (2) riverine desert (4.5%); (3) shield desert (23%); (4) desert clay plains (13%); (5) stony desert (12.5%); and (6) sand desert (32%). The riverine desert is the lowlands of the

Darling River drainage basin in the southeast, desert clay plains are the Vertisols in the northeast, stony deserts are the stony tablelands with silcrete gibbers (rocks) overlying deep clay located in the south and southeast, and the sand deserts are the great sand plains in the east central and west central part of the continent. Mountain and piedmont deserts, as well as the shield (granitic outcrops) desert, are dispersed across the arid zone.

Wind erosion is not a major problem on the aeolian sand plains which cover about one-third of the Australian arid zone. One reason is the relatively high clay content (10—20%) of about half of the sand and the other is the good vegetative cover of spinifex grass in most of the sandy areas. Where stabilized dunes do occur, they are about 10—30 m high and of the longitudinal (seif) type. Erosion by wind, if any, is generally confined to the crests of the dunes.

Vegetation

Spinifex (*Triodia* and *Plectrachne*) grasslands and mulga (*Acacia*) woodlands cover about 80% of the arid region. The remainder consists of saltbush—short grass—forb mixtures in the south and palatable grasses (e.g., *Eragrostis*) on the Vertisols of the northeast, plus some open eucalyptus (mallee) woodlands in the north and south. Spinifex dominates the sandy soils, especially, and serves to minimize wind erosion but is of little or no value as a forage crop (Perry, 1970). The acacias are moderately palatable and furnish important amounts of fodder during drought years. Most of the perennial and browse species provide little forage because they are rather unpalatable, leaving annuals as the major grazing resource during favorable periods.

SOILS

As Table 6.2 shows, soil associations dominated by Aridisols are the most extensive in the arid part of Australia. Together with the Entisol associations, they cover about 65% of the entire continent. In the western two-thirds of Australia, Entisols and Aridisols are intermingled. Arnhem Land and neighboring areas of the Northern Territory have a subhumid climate and a mixture of Entisols, Alfisols, and, probably, some Ultisols.

Alfisols

Alfisols occupy shrub woodlands where *Eucalyptus* and *Acacia* are extensive and *Callitris* and *Casuarina* are common in the wetter parts. On the map (Fig.6.1), Alfisols are shown to be concentrated in the southeast, south, and far west; they also are important components of the Entisol delineation in the northern, more humid sections of Queensland, the Northern Territory,

TABLE 6.2

Area of dominant soil orders of arid regions of Australia

Order	Area (km^2)	Percent of	
		arid region	continent
Alfisol	440,000	7.0	5.8
Aridisol	2,765,000	44.2	36.3
Entisol	2,285,000	36.6	30.0
Vertisol	760,000	12.2	10.0
Total	6,250,000	100.0	82.1

and Western Australia. In northern Australia, as in sub-Saharan Africa, Alfisols are the principal soils of the savannahs where the rainy period lasts for a few months during the summer and is followed by a long dry period. That kind of precipitation pattern produces an arid environment even though rainfall during the summer may amount to over 1,000 mm. In southern Australian woodlands, annual precipitation is much less than in the north but precipitation effectiveness is greater because more of it falls during the winter when evaporation is minimal.

Alfisols and Aridisols are included within the same Australian great soil groups or Primary Profile Form, in many instances. Solonized brown soils, for example, will be Alfisols if there is an argillic horizon present in a duplex (texture contrast) soil or Aridisols if an argillic horizon is absent and the profile form is gradational. This means that Australian great soil groups cannot be correlated completely with U.S. Comprehensive System orders; one Australian great soil group may include two or more U.S. orders.

AL1. Haplustalfs — Haplargids

These soils occur in four widely separated areas: along the west side of the Great Dividing Range on the east coast; around Adelaide and to the north; in far west Australia; and northeast of Perth. The topography is gently sloping to rolling in all four areas. The vegetative cover is savannah woodland (*Eucalyptus, Callitris, Casuarina*) with a good cover of xerophytic grasses (*Aristida, Stipa, Chloris, Eragrostis*). The climate ranges from arid to subhumid and the influence of past climates on present soil properties is considerable.

Haplustalfs belong to the red-brown earth great soil group in the Australian classification and to the Duplex Primary Profile Form in Northcote's key (Northcote, 1971). Duplex soils have texture-contrast profiles: subsoil texture is much finer than surface-soil texture and the boundary between the two is distinct. Surface texture is loamy sand to clay loam, subsoils are clay loam or clay, subsoil structure is prismatic to blocky, soil reaction varies from slightly acid to neutral, and the argillic horizon usually contains segregations of calcium carbonate in the lower part. Surface soils are gray-brown to red-brown; subsoils are reddish brown to red. Base saturation is high throughout the profile. Organic-matter content is moderate. The solum (A and B horizons) is 75—125 cm thick. Illite and kaolinite are the dominant clay minerals. Black manganiferous nodules are often present in the B horizon.

Haplargids of this association are in the red earths great soil group and have a Gradational Primary Profile Form in which there is a clear to gradual boundary between the coarser-textured A horizon and the finer-textured B horizon. In the normal soil, the surface horizon has colors ranging from grayish brown to red-brown, textures from loamy sand to sandy clay loam, and a massive and hard structure when dry. The argillic B horizon is highly porous, brittle to hard when dry, very friable when moist, the texture is sandy clay loam to clay, and the color is red to dark red. There may be a few black manganiferous nodules in the subsoil. Total depth of soil over weathered rock or sediments is one or more meters. Kaolinite and illite are the dominant clay minerals and the organic-matter content is moderate. The soil is calcareous in the lower subsoil. According to Stace et al. (1968), only minor clay illuviation (weak soil development) has occurred in this soil; textural variations are due more to alluvial deposition effects than to clay movement within the profile.

Ustorthents (earthy sands) are minor members of this association.

AL2. Haplustalfs — Calciorthids — Salorthids

As in the previous association, the climate ranges from arid to subhumid in the southern part of the continent where this association occurs. Salinity is a major problem in the wetter areas when the tree and shrub cover is removed during attempts to increase forage production or to cultivate the land. Replacement of the deep-rooted trees and shrubs by shallow-rooted grasses or crops leads to a rise in the water table and the development of seeps where the water reaches the surface. Since many of the Haplustalfs in this association have saline subsoils, the seeps often are saline. Vegetative cover on the flat to gently sloping topography is a low woodland of *Eucalyptus*, *Callitris*, *Casuarina*, *Heterodendron*, and *Acacia* with some *Atriplex* and *Kochia* shrub understory and grasses such as *Stipa*.

Haplustalfs of this association are the more developed of the solonized brown soils. They belong to the Duplex Primary Profile Form. Solonized brown soils formerly were known as mallee soils because of their eucalyptus trees; they are also called mallisols now, for the same reason. The soils have large amounts of calcareous material, with a calcic horizon commonly coinciding with the argillic horizon or beginning in the lower part of that horizon. Surface soils are gray-brown to red-brown sand to loam and are neutral to alkaline. Subsoils are less red, have finer textures of sandy clay loam to clay, and are more alkaline. Subsoil color is largely dependent upon the amount of calcium carbonate present and whether the carbonate is in nodules or is disseminated throughout the soil. Subsoils frequently are saline. Base saturation is high in all horizons and organic-matter content is moderate to high in the surface. Some manganese concretions may be found in the subsoil. Mallee soils may contain significant amounts of water-soluble boron (Penman, 1966). Many of these soils probably are Haplargids rather than Haplustalfs.

Calciorthids are the less-developed solonized brown soils and are classified as belonging to the Gradational Primary Profile Form. They are similar to the Haplustalfs but do not have an argillic horizon; clay content increases gradually with depth. The calcic horizon frequently is less well-defined in the Calciorthids than in the Haplustalfs of the solonized brown soils great soil group.

Salorthids are the Australian Solonchaks having a Uniform Primary Profile Form. They occur naturally on the lower slopes and in the basins in the solonized brown soils areas but also are man-made as the result of clearing trees and shrubs and of applying excess amounts of irrigation water. Salorthids vary widely in characteristics, but most of them have a medium or fine textured surface soil, a fine-textured subsoil, and a water table within 1 m of the surface. The source of the water is lateral movement from surrounding uplands. The salt comes from saline substrata, again in the uplands. Soils frequently have a saline crust during the dry season, and are calcareous at all depths. Groundwaters are salty. Vegetative cover is sparse, consisting mainly of *Atriplex* saltbushes.

There probably are some Haplargids within this delineation as well as Ustorthents, Torriorthents, and Ustifluvents.

Aridisols

Several of the Australian great soil groups include Aridisols of various kinds. Haplargids, for example, are found among the red earths, red-brown earths, calcareous red earths, siliceous sands, and desert loams. Aridisols may be of the Uniform, Gradational, or Duplex Primary Profile Form since it is aridity and the degree of development which determines whether a soil belongs in the Aridisol order rather than the distribution of clay. As with all Australian arid-zone soils, attempts at making a morphological or genetic classification are fraught with difficulties because it is hard to decide whether soil properties are carried over from an early humid climate or are a response to the present arid climate. One thing can be said about nearly all Aridisols in Australia: the dominant color is some shade of red, and most of the reds are brighter than in the reddish soils of northern hemisphere arid zones.

AR1. Durargids — Salorthids — Haplargids

Western Australia is the principal locale for this association of soils, the major constituents of which are red and brown hardpan soils and the saline loams of long, narrow depressions in the landscape. Precipitation is very low, on the order of 200—250 mm, and the plant cover is largely mulga (*Acacia*) which is a valuable source of browse during droughts. The topography is undulating.

Durargids are the red and brown hardpan soils having a Uniform or Duplex Primary Profile Form above the duripan. The duripan is the result of silica cementation of a layer of clay deposition and has the hardness of rock. It is reddish brown to red, has a vesicular porosity, and varies in thickness from several centimeters to many meters. Laminar layers of varying thicknesses are typically present in the pan and there are some surface coatings and accumulations of black manganiferous deposits. Above the duripan, which lies at a depth of 70—170 cm, the soil is reddish brown or red, and has a sandy loam to sandy clay surface underlain by an argillic horizon. There often is a desert pavement of siliceous gibbers and the profile itself may be stony. Soils are neutral to strongly acid in reaction and may be slightly calcareous immediately above the pan. Kaolinite and illite are the major clay minerals. Organic-matter content is low. Current thinking is that the duripan is a product of development under the present arid climate rather than being inherited from an earlier humid climate (Stace et al., 1968).

Salorthids (Solonchaks, Uniform Primary Profile Form) are found in former river valleys where a water table is relatively close to the surface as the result of seepage from surrounding uplands. The soils are somewhat stratified, medium to fine textured, saline throughout the profile, and frequently have a salt crust on the surface. The topography is undulating. Plant cover is sparse and consists mainly of salt-tolerant *Kochia* and *Atriplex*.

Haplargids of this association are soils of the red earths great soil group and have a Gradational Primary Profile Form. Red earths occur in a variety of rainfall zones, ranging from about 250 mm to 2,500 mm; therefore, only some of them qualify as Aridisols. The remainder resemble Alfisols more than any other U.S. order; many of them contain plinthite. Haplargids are in the 250—500 mm rainfall zone where the dominant shrub is *Acacia*. They are predominantly sandy in texture, but frequently have clay subsoils, are porous, red-brown to red in color, have massive structure and a weakly to moderately developed argillic horizon, and are hard when dry and friable when moist. The soils are acid to slightly alkaline in reaction, and may be calcareous in the lower part of the profile. They vary in depth from 1 m to several meters over rock. Kaolinite and illite are the principal clay minerals.

Other great groups in this association include Torriorthents and Torrifluvents.

AR2. Haplargids — Calciorthids — Salorthids

Two areas of this association are delineated on the soil map, one in northwestern New South Wales and southwestern Queensland and the other in southern Western Australia. Nearly all of the two areas consists of the calcareous red earths great soil group of Uniform and Gradational Primary

Profile Forms. There are minor inclusions of saline loams and gray-brown and red calcareous soils. Small areas of calcareous red earths occur fairly commonly in other parts of the arid regions. These soils are notable for their relatively high content of sesquioxides in the clay fraction. In general, the soils occur on undulating plains and pediments consisting of deposits of red materials eroded from older lateritic land surfaces. Annual rainfall is between 250 and 400 mm. Acacia (mulga) shrubland is the dominant plant formation.

Haplargids are the calcareous red earths which fit into the Gradational Primary Profile Form and usually have a reddish-brown loamy sand or sandy loam overlying a sandy clay loam or sandy clay argillic horizon. A massive structure prevails in the A and B horizons and both horizons are hard when dry, friable when moist, and highly permeable. Soil reaction changes from acid or neutral in the surface to alkaline in the subsoil. A highly calcareous horizon of segregated carbonate appears at a depth as shallow as 70 cm or as deep as two meters. The soils are deep to very deep. Gypsum may be present in the deep horizons. A few ferromanganiferous nodules sometimes show up in the B horizon. Organic-matter content is moderate to low. The carbonate source may be the parent material or atmospheric dusts or both. Illite and kaolinite are the principal clay minerals.

Calciorthids are members of two Australian great soil groups occurring in this association, one extensive and one inextensive. Calcareous red earths of the Uniform Primary Profile Form are extensive; the deeper members of the gray-brown and red calcareous soils, also having a Uniform Primary Profile Form, are inextensive. The top of the calcic horizon is at 30—40 cm in both groups. Surface soil in the calcareous red earths is a reddish yellow or yellowish red whereas the subsoils are red, changing to pink on the calcic horizon. In the gray-brown and red calcareous soils, the surface and subsoil is a reddish brown. Gypsum is present in the lower subsoils of both soil groups. The dominant clay minerals are illite and kaolinite. Organic-matter content is low to moderate. In the calcareous red earths, the surface horizon is usually free of carbonates whereas it is calcareous in the gray-brown and red calcareous soils. Both soils are alkaline throughout the profile. Gray-brown and red calcareous soils are 80—100 cm deep over rock; calcareous red earths are 1.5—2 m deep over remnants of old lateritic surfaces.

Salorthids belong to the Solonchak great soil group, called saline loams on the Stace et al. (1968) map, and the Uniform Primary Profile Form. Salorthids and Torriorthents are both represented among the saline loams; the presence or absence of a high water table is the differentiating feature. Saline water tables occurring at 1—2 m are responsible for development of Salorthids, which generally have a salt crust on the surface and are saline at all depths. Man-made Salorthids have become a major problem in southwestern Western Australia, as it has in New South Wales and South Australia, during the last several years. Grazing and cultivation pressure have led to removal of the native vegetation and a consequent increase in lateral seepage of groundwater.

AR3. Haplargids — Durargids — Haplustolls

Red earths, which constitute nearly all of the soils in this association, are scattered across the arid zones of Australia in small and large bodies. They also are present in subhumid eastern Australia where they are being used increasingly for crop production. Red earths occur on a variety of landscapes as well as in a wide range of climates. Land surfaces may be very old to young, stream levees, undulating plains, gently sloping pediments, hill slopes, or mesas. The one thing all the parent materials have in common is a rather large amount of silica. The other major great soil groups are the red and brown hardpan soils and lithosols. Plant cover consists of *Acacia* and *Eucalyptus* trees and *Triodia*, *Aristida*, *Stipa*, and *Danthonia* grasses.

Haplargids are the red earths (Gradational Primary Profile Form) of the arid region but the delineations on the map (Fig.6.1) include red earths in northern Queensland and the northern part of the Northern Territory which are Rhodustalfs or Haplustalfs rather than Haplargids. As in hot tropical Africa, Alfisols which are mainly distinguished from Argids by the degree of base saturation are really arid-region soils when they occur in climatic zones in which the precipitation is concentrated in the summer period, as is the case in northern Australia. Red earths have a massive structure, are predominantly sandy in texture and porous, and have a minimally developed argillic horizon. Soil reaction is acid to slightly alkaline, base saturation is high, and the soils are calcareous in the lower subsoil. Kaolinite and illite are the dominant clay minerals.

Durargids are red and brown hardpan soils having a Duplex Primary Profile Form above the duripan. Silica cementation of a layer of illuviated clay is responsible for duripan formation at depths of 70—170 cm. The duripan may be several centimeters or several meters thick and is very hard. The overlying soil and the duripan are reddish brown to red. Stones may be present in the sandy loam to sandy clay above the duripan and a desert pavement of siliceous gibbers is customarily present. Organic-matter content is low and the dominant clay minerals are kaolinite and illite.

Haplustolls are in the lithosol great soil group and have a Uniform Primary Profile Form. They have a high organic-matter content but no argillic horizon, are very dark gray-brown throughout the solum, and frequently are stony or gravelly. Textures range from sand to clay loam. The topography is hilly and the vegetation is eucalyptus and some grasses. Base saturation is moderate to high and the subsoils may or may not be calcareous. Rock underlies the soils at 50 cm or less and rock outcrops may appear on the surface.

Other great groups of local significance are Ustorthents, Torriorthents, and Chromusterts.

AR4. Haplargids — Torriorthents

Red siliceous sands are the major soils in this association. They are widespread in the Simpson, Victoria, Gibson, and Great Sandy deserts. There are minor inclusions of Entisols of the yellow earthy sands and Aridisols of the desert loams great soil groups. Red siliceous sands themselves are Haplargids although the sands are classified as members of the Uniform Primary Profile Form. The argillic horizon required for the Argid suborder represents a moderate accumulation of illuviated clay in a very sandy matrix and does not meet the criteria for a Gradational Primary Profile Form. Spinifex grass (*Triodia*) is the principal plant growing on the sand plains and low dunes of this association. Plant cover is sufficient to make wind erosion a minor problem except on the tops of some of the dunes. Salts are absent.

Haplargids in this association are deep red quartzose sands of uniform color and loose consistence even in the argillic horizon. Organic-matter content is generally low in the surface and very low in the subsoil, the soils are slightly acid to slightly alkaline in pH, and carbonates are absent to a depth of 1.5 m or more. Kaolinite and illite are the dominant clay minerals.

Torriorthents include the red siliceous sands which lack an argillic horizon and the yellow earthy sands, both of the Uniform Primary Profile Form. Profiles are uniformly loamy very fine sands or sandy loams and they range from 1 m to a few meters in depth. While the red siliceous sands are loose, the yellow earthy sands — also siliceous — have a coherent character which gives them the appearance of being finer textured than they actually are. Some of the earthy sands contain sesquioxide nodules, and large areas of low tablelands and sand plains in Western Australia are underlain at 1 m or less with either nodules or laterite crusts. There is little organic matter in these soils, with a slight accumulation in the surface, generally. Surface color is somewhat darker than the remainder of the profile, which is red or yellow. Kaolinite is the dominant clay mineral despite the siliceous character of the sands. Red siliceous sands are neutral to alkaline in reaction; yellow earthy sands are moderately to strongly acid.

Inclusions of desert loams in this association are classified as Haplargids. There are some Torripsamments among the earthy sands.

AR5. Haplargids — Haplustalfs — Haplustolls

Desert loams are the most extensive soils of this association, followed by solonized brown soils, red earths, and lithosols. They are concentrated in South Australia, New South Wales, and Queensland. All of the aforementioned great soil groups except the lithosols have argillic horizons. Desert loams have the Duplex Primary Profile Form; solonized brown soils and red earths are Gradational; lithosols are Uniform. The different Primary Profile Forms mean that Duplex soils have a distinct argillic horizon, Gradational soils have an indistinct argillic horizon or may have no such horizon, and Uniform soils do not have an argillic horizon. The soils in this delineation occur on wide alluvial plains and on stony tablelands in regions with less than 250 mm of precipitation. Textures are always loamy in the surface. Kaolinite and illite are the principal clay minerals. Saltbush (*Atriplex*, *Kochia*, *Bassia*) and *Nitraria* are the extensive shrubs; *Danthonia*, *Stipa*, *Eragrostis*, and *Aristida* are the important perennial grasses.

Haplargids of this association include the well-developed desert loams and the moderately developed red earths. Typically, desert loams have a thin loam A horizon overlying a distinct clay-textured argillic horizon. Surface colors range from reddish brown to red, subsoil color is red, and soil reaction is usually neutral to moderately alkaline. Surface soil has a massive structure with a vesicular layer about 1 cm in thickness at the top. B horizons have a blocky structure, are friable when moist, contain some gypsum, and are moderately to strongly saline, with 20% or more of exchangeable sodium in the lower part. The soils are deep and calcareous at all depths except, occasionally, the surface. Desert pavements are common in some areas but absent in others. Organic-matter content is very low to low. Haplargids belonging to the red earths great soil group are predominantly sandy in texture and have a weakly to moderately developed argillic horizon. They are moderately deep to deep and may be calcareous in the lower horizons.

Haplustalfs represent the more developed of the solonized brown soils (those having a Duplex Primary Profile Form). They have a calcic horizon in or below the argillic horizon, are noncalcareous on the surface, have moderate to high organic-matter contents. Subsoils frequently are saline and may contain manganiferous concretions. Surface-soil color is gray-brown to reddish brown and surface textures range from sand to loam. Subsoils are less red and have sandy clay loam to clay textures. Base saturation is high in all horizons.

Haplustolls are lithosols of the shallow loamy soils subdivision. They have a mollic epipedon but no argillic horizon, base saturation is high, there may or may not be a calcareous layer in the profile, rock outcrops are common, and the soils may be stony or gravelly. Soil color is very dark gray-brown. They occur on moderate to steep slopes, usually. Rock underlies the soil at 50 cm or less.

Torriorthents are minor soils in this association.

AR6. Calciorthids — Torriorthents

Soils of this association are found mainly in the south: on the Nullabor Plain and in central and eastern South Australia. They belong to the gray-brown and red calcareous soils great soil group and the Uniform Primary Profile Form. The soils are developed in place from calcareous sedimentary rocks which underlie them at shallow to moderate depths. The topography is level and the vegetation consists of mallee (*Eucalyptus*), mulga (*Acacia*), saltbush (*Atriplex*, *Kochia*, *Bassia*), and grasses (*Aristida*, *Stipa*, *Triodia*). The main clay minerals are illite and kaolinite.

Calciorthids of the gray-brown and red calcareous soils (Uniform Primary Profile Form) are the deeper ones having a calcic horizon; the shallow ones are Torriorthents. In the Calciorthids, the top of the calcic horizon is at 30—40 cm but the soils are calcareous in all horizons. Surface and subsoil color is gray-brown to red and soil structure is massive. Often there is a desert pavement on the surface. Textures are loam or clay loam and there is little or no differentiation with depth. Gypsum and some manganiferous nodules may be present in the lower part of the profile, organic-matter content is moderate, and the soils are about 80—100 cm deep over rock. They are non-saline.

Torriorthents are the shallow members of the gray-brown and red calcareous soils which do not have a calcic horizon. In all other characteristics, they are similar to the Calciorthids in this great soil group. Fragments of limestone frequently are present in the solum and may be lying on top of the soil. Colors range from gray-brown to reddish brown and the uniform texture is usually loam or clay loam. Organic-matter content is moderate to low.

There are inclusions of Haplargids (desert loams, red earths), Salorthids (Solonchaks), Ustorthent (Solonchaks), and Haplustalf (red-brown earths) great groups.

Entisols

E1. Ustorthents — Haplustalfs

Shallow soils are found in a long arc around western, northern and north-eastern Australia, in the Macdonnell Ranges of central Australia, in the Gawler Ranges of South Australia, and elsewhere in smaller areas. In the north and east, the delineations extend into subhumid and humid regions because soil conditions are similar even though the vegetation is a denser woodland in the wetter section. Soils are shallow because of erosion and they occur, therefore, on slopes and tops of hills and mountains. The plant cover is sparse acacia and eucalyptus shrubland in the drier areas and tropical woodland in the wetter northern areas with an understory of unpalatable grasses. Illite and kaolinite are the principal clay minerals.

Ustorthents, in this case, are predominantly lithosols of the Uniform Primary Profile Form, with no horizon development other than an A horizon of organic-matter accumulation. Textures usually are sandy loams, loams, and clay loams; rocks are present on the surface; soils usually are stony or gravelly; and bedrock is found at less than 50 cm, generally. The soils may be calcareous or noncalcareous. Soil colors are gray-brown to yellow-brown. Organic-matter content is moderate to high. In the more humid regions of the south and east, some of the lithosols are dark enough to qualify as Haplustolls. The yellow earthy sands in this association are Ustorthents or Torriorthents, depending upon the aridity of the climate.

Haplustalfs are of minor extent among the lithosols in the arid and semi-arid climates but are important in the subhumid and humid climates of the tropical woodlands on the northern fringe of this association. Alfisols are found in several of the Australian great soil groups in the tropical climatic zone, including the red-brown earths, yellow earths, noncalcic brown soils, and terra rossa soils and the Uniform, Gradational, and Duplex Primary

Profile Forms. Haplustalfs probably are the most widespread of the Alfisols. They have an argillic horizon at 10—30 cm, may or may not be calcareous in the lower part of that horizon, have gray-brown to red-brown loamy sand to clay loam surface soils and reddish-brown to red finer-textured (frequently clay) subsoils, are acid in the surface, and have a base saturation greater than 50%.

Among the variations in the Alfisols of this association are Paleustalfs (limestone within 1 m of the surface), Natrustalfs (a natric horizon instead of an argillic horizon) and Rhodustalfs (very red argillic horizon). There also are some Haplargids (red earths) and Salorthids consisting of saline clays along the northern coast.

E2. Torripsamments — Ustorthents — Haplargids

The dominant soils in this association are the earthy sands, usually red in color but sometimes yellow. There also are significant areas of lithosols, red earths, and earthy sands with ironstone gravels, plus occurrences of solonchaks. The landscape consists of low tablelands and plains and the vegetation is mainly spinifex grass (*Triodia* and *Plectrachne*). Illite and kaolinite are the main clay minerals.

Torripsamments probably include most of the earthy sands but they verge upon being Ustorthents in the areas peripheral to the central arid zone. Soil profiles are similar for the two great groups. In both, the profiles are of the Uniform Primary Profile Form and are coarse-textured sands, clayey sands, or sandy loams. The surface soil is a little darker than the subsoil. There is no accumulation of soluble salts, carbonates, or gypsum in the profile and soil depth is commonly about 1 m over iron-oxide nodules or indurated material. Soil reaction is moderately to strongly acid. Iron-oxide (ironstone) gravels are present in considerable quantities in earthy soils of large sections of the Gibson Desert and, to a lesser degree, in the Great Sandy Desert. Lithosols included in the delineation also belong to the Torriorthent or *Ustorthent* great group. Lithosolic Orthents are shallow, usually stony or gravelly, have uniform textural profiles, and are mostly noncalcareous.

Haplargids are the argillic Aridisols of the red earths great soil group and the Gradational Primary Profile Form, with a clear to gradual boundary between the A and B horizons. They are massive and hard when dry, predominantly sandy textured (loamy sand to sandy clay loam), porous, red-brown to red in color, and have weak profile differentiation. Soil reaction is slightly acid to neutral, calcium carbonate is usually present in the lower part of the subsoil, and base saturation is high. Soils are deep over weathered rock or sediments. Nodules of manganiferous or ferruginous composition are usually present in variable amounts in the subsoil. Organic-matter content is moderate.

Salorthids are found in some of the ephemeral lake beds as well as along coastal beaches and in estuaries.

Vertisols

A very large part of arid Australia is covered with deep cracking clays of the Vertisol order. Nearly all are located in the east and northeast on broad level river plains and gently undulating upland plains. About the only property they have in common is that of being cracking clays; all other properties vary widely from place to place, including color, mottling, carbonate and gypsum content, salinity, exchangeable sodium, manganese and iron segregations, pH, and permeability. Annual precipitation ranges from 250 to 750 mm. Mitchell grass (*Astrebla* spp.) is a major component of the native plant cover on the northern Vertisols but heavy grazing pressure has caused deterioration of the plant communities and partial replacement of Mitchell grass by *Eragrostis* and annual forbs and herbs. On the southern Vertisols, the vegetation consists primarily of eucalyptus, grasses such as *Eragrostis*, *Panicum*, and *Dichanthium*, various forbs, and saltbush (*Atriplex*). Montmorillonitic clays are present in amounts large enough to produce much expansion and contraction of the soils after wetting and drying. Sometimes montmorillonites constitute a majority of the clays, other times they may be equal to or less than illites and kaolinites. Vertisols are the only soils in arid Australia which contain important quantities of montmorillonite. The typical hummocky relief of Vertisols known as *gilgai* has been studied more in Australia than anywhere else.

V. Chromusterts — Pellusterts — Haplargids

The great majority of soils are Vertisols of one kind or another. Other soils are mostly red earths belonging to the Haplargid great group. Chromusterts are the brown and red Vertisols; Pellusterts are the gray Vertisols.

Vertisols are the gray, brown, and red clays having little or no profile development and a Uniform Primary Profile Form. They are deep to very deep, contain 50—80% clay, may be calcareous and gypsiferous in the subsoils, frequently are saline below 50 cm, sometimes have mottled subsoils when drainage is poor, may or may not contain considerable exchangeable sodium in the subsoil, and have a blocky structure. High exchangeable sodium levels are commoner in the Vertisols in the southeast than in the other Vertisols. Soil reaction is usually slightly acid to moderately alkaline and base saturation is high. A few soils are moderately or strongly acid and some have the highly unusual condition of an alkaline and calcareous surface soil overlying a strongly acid subsoil. Dehydration cracks are generally 2—10 cm wide and as much as 130 cm deep. The gilgai relief is variable, ranging from small mounds a meter across and 25 cm high to hummocks several meters in width and 0.5—2 m in height. Some Vertisols have no gilgai features. Water tables within 2 m of the surface occur along some of the watercourses. Parent material is unconsolidated clay alluvium or weathered rocks which produce high-clay soils, such as on the Barkly Tableland in the Northern Territory and Queensland.

Haplargids are the red earths having a Gradational Primary Profile Form. Surface soils vary from loamy sand to sandy clay loam, and are massive and hard when dry. Colors are gray-brown to reddish brown. Subsoils are sandy loam to clay, red in color, and are massive and hard when dry and friable when moist. Most of these Haplargids are slightly to moderately acid and may be calcareous in the lower subsoil. Manganiferous and ferruginous nodules are of variable occurrence. The soils are deep or very deep and the dominant clay minerals are kaolinite and illite.

Durargids of the red and brown hardpan soils and Torrifluvents are other great groups in this association.

SOIL DEVELOPMENT

Australian arid-zone soils appear to be in slow transition from typical soils of the humid tropics to those representative of the majority of the world's arid regions. Except for places where calcareous rocks are dominant, such as the Nullabor Plain, the soils retain at least vestiges of characteristics developed during the millions of years when the climate was wet. Thus, soil reaction commonly is acidic, silcrete and laterite boulders (gibber) are widespread, strong red colors are common, and illite and kaolinite are the dominant clay minerals. Some of the soils also have organic-matter contents which are unusually high for arid-region soils, frequently attaining levels of 4% or more.

During the last one million years when the climate became increasingly arid, a gradual change appears to have been underway. Currently, the bewildering variety of soils found in the arid regions defies attempts at a genetic classification, which has led to the Northcote Key placing major emphasis on textural profile forms rather than on genetic characteristics. With such a variety of soils occurring in what is now an arid region, it is easy to understand why many Australian soil scientists contend that Australian arid-region soils are proof that there is no such thing as a typical arid-region soil-forming process.

The extreme soil variation probably is due to the widespread redistribution by water erosion, in particular, of soil materials during the Quaternary period and to the slower rate of development associated with increasing aridity. In their book, *A Handbook of Australian Soils*, Stace et al. (1968) make numerous references to sedimentary layering in the arid-region soils. Subsoil layers frequently have properties resembling those of soils developed under humid conditions, which would be expected if little change had occurred after erosion denuded the upland areas and deposited the soil material in lower-lying areas. If this hypothesis is correct, it will be a long time, if ever, before Australian arid-region soils resemble those of the northern hemisphere, but the fundamental soil-forming processes lead in that direction. The change occurring in Australian soils may be similar to what might be expected if the Alfisols of sub-Saharan Africa were exposed to the more arid climate of the Sahara.

REFERENCES

Mabbutt, J.A., 1969. Landforms of arid Australia. In: R.O. Slatyer and R.A. Perry (Editors), *Arid Lands of Australia.* Australian National University Press, Canberra, A.C.T., pp.11—32.

Meigs, P., 1953. World distribution of arid and semi-arid homoclimates. In: UNESCO, *Reviews of Research on Arid Zone Hydrology. Arid Zone Res.*, I: 203—210.

Northcote, K.H., 1971. *A Factual Key for the Recognition of Australian Soils.* Rellim Technical Publications, Glenside, S. A., 123 pp.

Perry, R.A., 1970. Productivity of arid Australia. In: H.E. Dregne (Editor), *Arid Lands in Transition. Am. Assoc. Adv. Sci., Publ.*, No. 90: 303—316.

Penman, F., 1966. Slow reclamation by tile drainage of sodic soils containing boron. In: *Int. Comm. on Irrigation and Drainage, Congr., 6th*, Question 19, R. 10, pp.19.113—19.121.

Stace, H.C.T., Hubble, G.D., Brewer, R., Northcote, K.H., Sleeman, J.R., Mulcahy, M.J. and Hallsworth, E.G., 1968. *A Handbook of Australian Soils.* Rellim Technical Publications, Glenside, S. A., 435 pp.

U.S. Department of Commerce, 1969. *Climates of the World.* Environmental Data Service, Environmental Science Services Administration, U.S. Department of Commerce, Washington, D.C., 28 pp.

Chapter 7

NORTH AMERICA

INTRODUCTION

The arid region of North America forms a large block of land lying west of 98°W and between 20° and 54°N, in addition to one area in southern Mexico (Fig.7.1). Several major rivers in the region such as the Columbia, Colorado, Missouri, and Rio Grande (Rio Bravo in Mexico), plus numerous smaller ones, have their source in the mountains within or bordering on the arid regions. Agriculture (dryland farming, irrigated cropland, and livestock raising) and mining are the major industries. Urban centers, exemplified by the Juarez—El Paso metropolitan complex, are oases surrounded by a sparsely populated hinterland. East of the Rocky Mountains of Canada and the United States lies the semiarid and subhumid Great Plains where most of the world's exportable supplies of wheat are grown. Not shown on the map are the cold deserts of northern Canada and Greenland. Insufficient information is available to classify the soils of the cold arid regions.

Aridisols and Mollisols dominate the region; Entisols are important inclusions in all the soil areas; and Alfisols and Vertisols are minor soils compared to the other orders but are dominant in some areas (Fig.7.2). Information on soils came largely from the North America map (FAO/UNESCO, 1972) of the Soil Map of the World project; a regional publication of the western states agricultural experiment stations and the Soil Conservation Service (U.S. Department of Agriculture, 1964); a map compiled by Aandahl (1972); and a map compiled by the Secretaría de Recursos Hidráulicos (1970).

ENVIRONMENT

Climate

The North American arid land mass lies nearly entirely within the middle and northern latitudes. Along the west coast, a Mediterranean type of climate prevails (wet winters, dry summers; moderate temperature) whereas the Great Plains has a continental climate (precipitation throughout the year; marked seasonal temperature extremes). Temperature variations are less extreme in the highlands of Mexico than in the elevated plateaus of the United States and in the Great Plains. Snowfalls are heavy in the northern and eastern part of the arid regions (and in the high mountains), tapering off

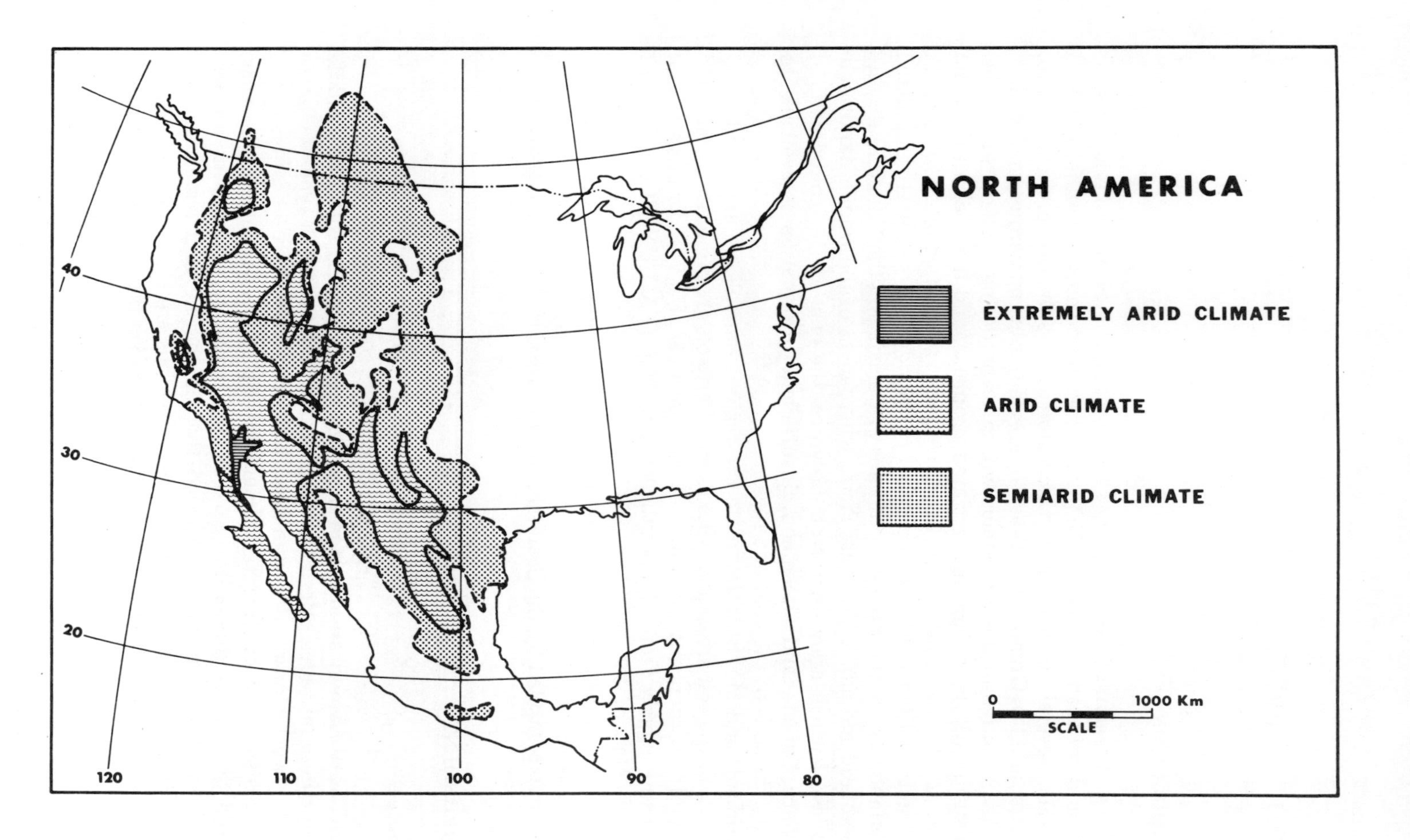

Fig.7.1. Arid regions of North America (after Meigs, 1953).

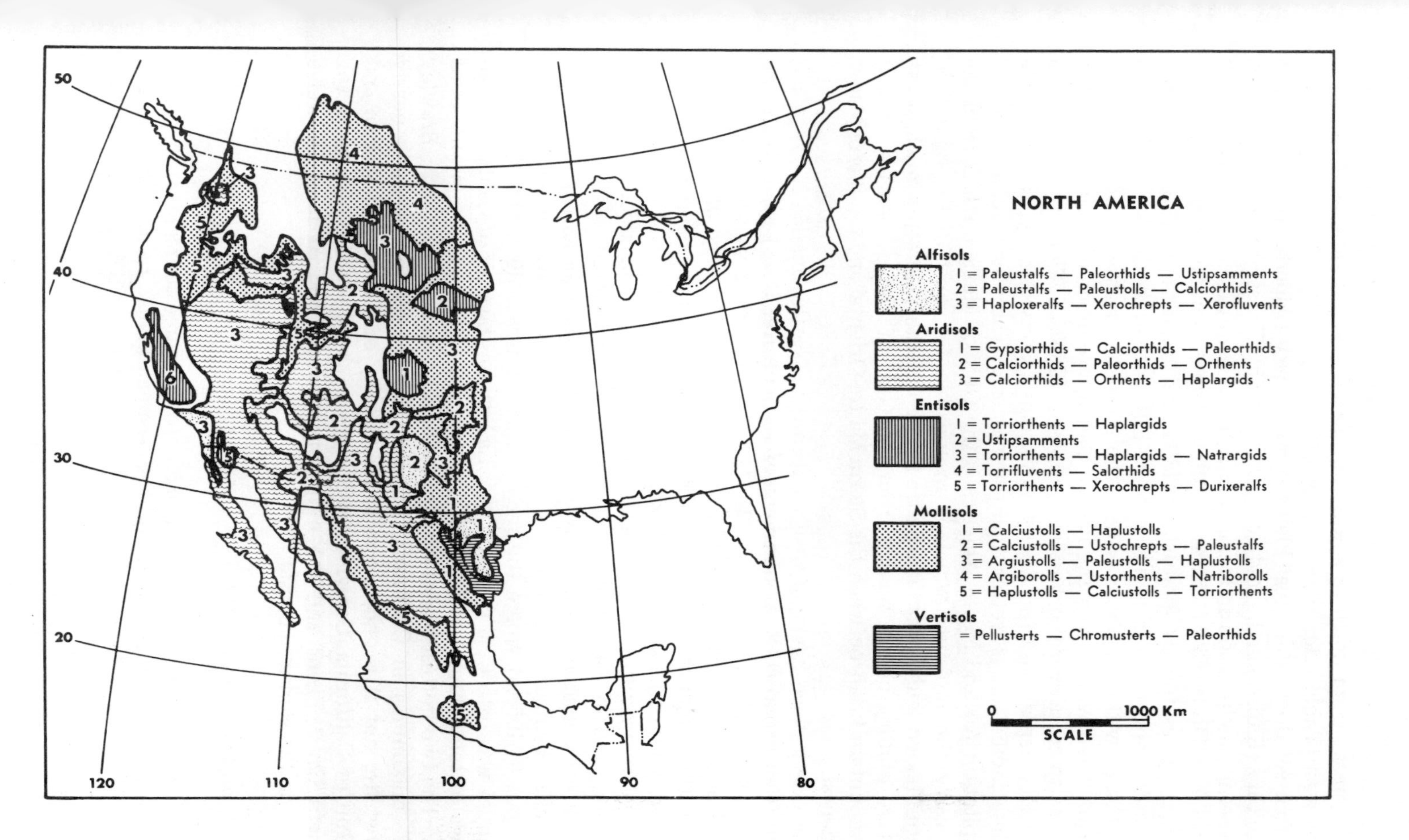

Fig. 7.2. Arid-region soils of North America. Polar soils not shown.

to little or none in the south and west. Generally speaking, annual temperatures increase from north to south and annual precipitation decreases from east to west (Table 7.1).

A transect down the Great Plains from Calgary (Alberta) on the north to Dodge City (Kansas) in the middle to Brownsville (Texas) on the south shows increases in precipitation and temperature in the same direction. An east—west transect from Dodge City (Kansas) to Albuquerque (New Mexico) to Las Vegas (Nevada) passes toward increasing aridity and higher annual temperatures.

Elevation above sea level has a strong influence on both precipitation and temperature within regions, as does the presence or absence of mountains. Moisture-laden winds from the Pacific Ocean lose much of their water when they rise to cross over the western mountains (Cascade, Sierra Nevada, Sierra Madre Occidental, and the Rocky Mountains). Consequently, the west side of these mountains is wetter than the east side. The other major source of precipitation for arid North America is the Gulf of Mexico. Winds from the Gulf carry the moisture which is responsible for most of the rainfall in eastern Mexico and Texas. The Sierra Madre Oriental intercepts much of the moisture coming across Mexico from the Gulf of Mexico, leaving an arid intermountain basin between the Sierra Madre Oriental and the Sierra Madre Occidental.

The driest place having a weather record in North America is Greenland Ranch (California) (U.S. Department of Agriculture, 1941) where the mean annual precipitation is 38 mm. In Mexico the driest weather station is located at La Paz (Baja California) with an annual average of 145 mm. In Canada, the minimum annual precipitation of 140 mm occurs at Resolute (Northwest Territories) in the cold desert at about 75°N.

Annual temperatures are highest along the Gulf of Mexico, the Gulf of California, and in the states bordering the lower Colorado River; they are lowest in the Canadian Great Plains and in the high plateaus among the Rocky Mountains. Temperature extremes are greatest for the stations in Table 7.1 at Calgary (Alberta) where the record maximum was 36°C and the record minimum was —45°C, a range of 81°. La Paz, on the Gulf of California, had a range of temperature of only 43° (a maximum of 42° and a minimum of —1°). The two hottest places are Blythe (50°C maximum) and Brawley (49°C maximum) in California. Frost is experienced occasionally nearly everywhere, but it is uncommon in the vicinity of the Gulf of Mexico and the Gulf of California. Growing seasons are as short as three months in the northern Great Plains and the mountain plateaus and as long as twelve months on the southern coasts.

Geology and geomorphology

Within the arid region of North America are three subregions: the Great Plains, intermountain plateaus, and the basin-and-range region. Between the

TABLE 7.1

Climatological data for six North American weather stations

Station	Elevation (m)	Precipitation (mm)													Average annual temperature (°C)
		J	F	M	A	M	J	J	A	S	O	N	D	annual	
Calgary (Canada)	1,082	13	13	20	25	58	79	64	58	38	18	18	15	419	4
Dodge City (U.S.A.)	787	15	18	30	46	81	76	58	61	38	36	15	13	487	13
Albuquerque (U.S.A.)	1,619	10	10	13	13	20	15	30	33	25	20	10	13	212	14
Las Vegas (U.S.A.)	659	13	10	10	5	2	0	13	13	8	5	8	10	97	19
Brownsville (U.S.A.)	5	36	38	25	41	61	76	43	71	127	89	33	43	683	23
Chihuahua (Mexico)	1,350	5	10	8	5	5	43	91	94	84	23	13	10	391	18

Source: U.S. Department of Commerce (1969).

arid subregions are the humid mountains where the headwaters of major rivers are located.

The Great Plains, on the east, consists of more or less parallel planes of sedimentary rocks (shale, sandstone, limestone) sloping gently eastward from the Rocky Mountains. In the High Plains of Texas and New Mexico, the land surface is almost flat but in most of the Great Plains the topography is gently undulating to rolling, with the main stream valleys having a dominantly east—west orientation. In the highly eroded badlands of Montana, North Dakota, and South Dakota, the topography is hilly. A cap of aeolian material overlies much of the Great Plains, and surficial loessial (windblown silt) deposits are important, particularly in Colorado, Nebraska, Kansas, Oklahoma, New Mexico, and Texas. The Nebraska Sand Hills covering a large part of central and western Nebraska are a product of wind erosion, and wind erosion is a continuing hazard on cultivated lands. Cold air masses sweeping across the plains from northern Canada and the occasional warm chinook winds are responsible for sudden and dramatic temperature changes in winter.

There are four major units in the intermountain-plateaus subregion: Columbia Plateau in Washington, Idaho, and Oregon; Colorado Plateau in Arizona, Colorado, New Mexico, and Utah; Wyoming Basin in Wyoming; and the Mexican Highlands in Arizona, New Mexico, and the states from Chihuahua to Querétaro. Shale, sandstone, and limestone are the dominant underlying rocks. A mantle of aeolian material covers the basalt which underlies much of the Columbia Plateau. Coarse alluvial and colluvial deposits are common and there are fine-textured playas (closed basins) in the Mexican Highlands. The overall aspect in all the plateaus is planar but the local relief is rolling to hilly, in the main, and there are scattered mountains. The Colorado Plateau is dissected by many deep canyons. Elevations in the intermountain-plateaus subregion range from 150 to 2,200 m.

The basin-and-range subregion extends south from Oregon and lies between the Rocky Mountains and the Cascade—Sierra Nevada mountains in the United States and west of the Sierra Madre Occidental in Mexico. It consists of a repetitive pattern of mountains and intervening basins, most of the latter being of the closed variety having no surface outlet to the Pacific Ocean. The mountains are composed mainly of old sedimentary rocks but many of them are igneous. Basins vary in size from a few to many kilometers across. Below the steep mountains slopes are coarse-textured alluvial and colluvial fans. The center of the basin is a fine-textured and often saline playa. Sand dunes frequently are present in the more arid areas where the vegetative cover is minimal. In Baja California, Sinaloa, and southern Sonora, the basin-and-range landscape changes gradually to a valley-and-range relief as more of the streams find an outlet to the sea.

Vegetation

In North America, the kind of shrub which is present indicates whether the climate is that of the cool or the hot arid regions. Sagebrush (*Artemisia*) is an indicator of the cool arid region occurring in the north and at higher elevations in the south. Creosotebush (*Larrea divaricata*) and mesquite (*Prosopis* spp.) are indicators of the hot arid regions at lower elevations in the south. Next to sagebrush, shadscale (*Atriplex confertifolia*) is the most important cool-region shrub. *Atriplex canescens* (fourwing saltbush) is a common shrub in the hot arid regions, along with cactus (*Cactaceae* spp.) and tarbush (*Flourensia cernua*).

The Great Plains is a grassland in which shrubs and trees are confined largely to valley bottoms, north slopes of hills, and shallow rocky soils. Temperatures decrease from south to north and precipitation decreases from east to west. In the north, the dominant grasses are *Agropyron*, *Stipa*, and *Bouteloua*, and shrubs are common only in the drier part of the plains. In the south, *Bouteloua*, *Buchloe*, *Aristida*, and *Andropogon* are the principal grasses, but shrubs and trees (*Prosopis* and *Quercus*) are widespread on upland sites where overgrazing has occurred.

Vegetation of the intermountain plateaus and the basin-and-range region varies from north to south, as it does in the Great Plains, but shrubs are a more important component of the plant ecosystem. *Artemisia* and *Atriplex* may be present in thick stands in the cool arid regions where the main grasses are *Agropyron* and *Festuca*. In the hot arid regions, perennial grasses (*Bouteloua* and *Hilaria*) are found among the *Larrea* and *Prosopis* in the wetter places but frequently are absent in the drier parts of Arizona, Sonora, Baja California, and California. *Larrea* is the dominant drought-tolerant shrub on upland sites where coarse-textured calcareous soils occur. *Prosopis* provides the principal shrub cover on sandy sites.

SOILS

Arid North America consists mainly of Aridisols and Mollisols (Table 7.2). Canadian arid regions are dominated by Mollisols; United States regions have about equal amounts of Aridisol and Mollisol associations; and Mexico has much more Aridisols than Mollisols. Approximately 2.8% of Canada has arid-region soils (not counting the cold deserts of the northern islands), 34% of the United States, and 49% of Mexico. Alfisols occur in three different climate zones: those in south Texas are in a hot summer rainfall area, those along the Texas—New Mexico border have a continental climate, and the Alfisol area in California and Baja California has a Mediterranean climate with moist winters.

TABLE 7.2

Area of dominant soil orders in arid regions of North America

Soil order	Area (km^2)	Percent of	
		arid region	continent
Alfisol	165,000	3.8	0.7
Aridisol	1,950,000	44.8	8.1
Entisol	350,000	8.0	1.4
Mollisol	1,790,000	41.1	7.4
Vertisol	100,000	2.3	0.4
Total	4,355,000	100.0	18.0

Alfisols

Typical Alfisols have an ochric (light-colored) epipedon, an argillic horizon and a moderate to high base saturation and are moist in the subsoil for at least three months during the warm part of the year. In North America, Alfisols are not extensive in the arid regions but they are extensive in the humid regions, including the humid mountain "islands" within the arid regions. The soil temperature regime for arid-region Alfisols is mesic, thermic, or hyperthermic. The soil moisture regime is mainly ustic but becomes xeric in the Mediterranean climate of the Pacific Coast. Clay minerals are mixed or are montmorillonitic.

AL1. Paleustalfs — Paleorthids — Ustipsamments

The part of Texas in which this association is found consists of a nearly level or undulating coastal plain sloping gradually toward the Gulf of Mexico. Elevations range from 10 up to 300 m. Annual precipitation varies from 500 to 750 mm, most of it coming in spring and summer. Rainfall distribution is erratic, droughts and floods are common occurrences. Summer temperatures are high and growing seasons are long. The original vegetation was a grassland or savannah type but long-continued grazing has been followed by an increase in shrubs and trees (*Prosopis*, *Quercus*, *Acacia*) to the point where the latter are now dominant. The remaining grasses include *Bothriochloa*, *Paspalum*, *Setaria*, *Buchloe*, *Chloris*, and *Hilaria*.

Paleustalfs have a moderately thick A horizon above the argillic B horizon which lies at 45—60 cm below the surface. Surface soils are reddish brown in color whereas subsoils are yellowish red or yellowish brown. Textures commonly are fine sandy loam on the surface and sandy clay loam in the argillic horizon although many Paleustalfs are finer textured in both the surface soil and the subsoil. Soil reaction is slightly to moderately acid in the surface and subsoil, but the lower subsoil in the medium and fine textured soils usually is calcareous. Base saturation is moderate to high. Organic-matter content approaches 1%. Practically all of the soils are at least moderately permeable and more than 1.5 m deep but some have a petrocalcic horizon at a depth of 0.8—1.5 m.

Paleorthids have a petrocalcic horizon at 20—50 cm, do not have an argillic horizon, have a moderate organic-matter content, are grayish brown in surface color, are calcareous throughout the profile, and occupy old level to gently sloping surfaces. Textures are medium, in the main, and permeability is good since cracks are frequent in the petrocalcic horizon. The surface soil contains angular caliche fragments.

Ustipsamments are deep loamy fine sands and fine sands showing little evidence of development. They are neutral to moderately acid in reaction, low in organic matter, brown to yellowish brown in color, and are found on gently undulating to rolling landscapes.

Among the other great groups are Haplustalfs, Camborthids, Torrerts, Ustifluvents, and Ustochrepts.

AL2. Paleustalfs — Paleustolls — Calciorthids

Alfisols, Mollisols, and Aridisols grade into one another in the southern High Plains of New Mexico and Texas. The plains are so flat that the greatest change in local elevation may be only the 20 m between the plain proper and the bottom of one of the thousands of closed depressions called playas. Annual precipitation varies from 300 mm in the southwest corner to 500 mm in the northeast. Peak-rainfall periods come at the end of May and August. The growing (frost-free) season amounts to between 200 and 230 days. Summers are hot and winters are cold. The vegetation is grassland, with *Bouteloua*, *Buchloe*, and *Andropogon* the principal grasses. *Prosopis* has infested the rangelands of the south and southwest. Shinnery oak (*Quercus havardii*) and mesquite (*Prosopis juliflora*) have invaded many of the eroded sandy soils.

Paleustalfs of the High Plains have a fine sand to fine sandy loam surface soil overlying a sandy clay loam argillic horizon at 35—70 cm. They are highly susceptible to wind erosion. Surface soils are brown and subsoils are yellowish red or red in color. Organic-matter content is low and the soils are neutral to slightly acid in reaction and are noncalcareous to a depth of 50 cm or more. Permeability is good.

Paleustolls have a dark-brown loam to clay loam surface texture, a reddish-brown clay loam argillic horizon at 17—23 cm, and a moderate organic-matter content, are slightly to moderately alkaline in reaction and commonly are calcareous below 1 m, and are moderately permeable.

Calciorthids are very shallow soils over thick layers of soft or hard caliche. They are pale brown to grayish brown in surface color, strongly calcareous to the surface, verge on having a mollic epipedon, and have surface textures ranging from fine sandy loam to clay loam. The calcic or petrocalcic horizon appears at 5—25 cm below the surface and is highly permeable to water and roots.

Other important great groups are Ustipsamments, Calciustolls, and Pellusterts.

AL3. Haploxeralfs — Xerochrepts — Xerofluvents

Alfisols are found on alluvial fans, terraces, and low mountains, Inceptisols on the rolling mountain foothills, and Entisols in the river valleys and higher mountains in this association extending from Los Angeles to northern Baja California. Annual precipitation over most of the area varies from 250 to 600 mm. Summers are dry and winters are wet. The growing season at the higher elevations may be as short as 180 days, whereas frosts are rare or unknown along the shore of the Pacific Ocean. The vegetation consists of a grass and shrub combination, with trees (*Pinus*) in the wetter and cooler subhumid mountains. *Bromus*, *Stipa*, and *Poa* are the dominant grasses; *Quercus* is the dominant shrub. In places where the shrub growth is dense and the vegetation type is referred to as chaparral, the principal shrubs are *Quercus*, *Ceanothus*, *Arctostaphylos*, and *Rhus*. The topography is hilly and steep except for the level flood plains and river terraces.

Haploxeralfs have a brown or reddish-brown sandy loam to clay loam surface soil which is 10—20 cm thick and overlies a reddish-brown sandy clay loam to clay loam argillic horizon. Organic-matter content is low, base saturation commonly exceeds 80%, surface-soil reaction is slightly acid, and calcium carbonate frequently is found in the lower part of the B horizon. Soil depth usually is less than 1 m over parent rock. Many profiles are stony.

Xerochrepts exhibit the incipient development typical of Inceptisols and are shallow to moderately deep over granitic and basic igneous bedrock. Surface soils are brown and subsoils are light brown. The profile texture is sandy loam to loam, soil reaction is neutral to slightly alkaline, and there may be calcium carbonate accumulations just above the granite. Stones are common in the profile.

Xerofluvents are the recent alluvial soils of the flood plains and river terraces where rivers deposit their sediments. Most of the soils are medium textured and deep, show little or no development, are low to moderate in organic matter, have a neutral to slightly acid reaction, may be calcareous in the subsoil, and are stratified.

Palexeralfs and Ustorthents are the main great groups of secondary importance.

Aridisols

A variety of soils has been developed under an aridic moisture regime in North America. Virtually all of them — except for some sand-dune soils —

have one characteristic in common: they are calcareous within 1 m of the surface. Additionally, the vast majority of the soils have a slightly acid (on the surface) to strongly alkaline reaction, with a pH of the saturated paste which seldom is below 6.0. Upland soil textures are dominantly sandy loam and loam but alluvial fans and mountain footslopes commonly are gravelly and stony. Soils in the depressions in closed basins are fine textured. River-valley flood plains vary widely in texture within short distances and the soils usually are stratified with layers of sand, silt, clay, and gravel. Clay minerals are of mixed types (montmorillonite, kaolinite, illite, hydrous oxides, chlorite, allophane) but montmorillonitic minerals predominate. Canada has a considerable amount of cold arid-region soils, none of which are shown in Fig.7.2 because their extent is unknown. The little information available on North American cold-desert soils indicates that they are concentrated above 70°N on Ellesmere Island and the Queen Elizabeth Islands of Canada and on the ice-free edges of Greenland. Similar soils undoubtedly occur on high peaks of the Rocky Mountains in Canada and the United States.

Soils known formerly as Gray Desert or Serozem soils are dominant in the cooler arid regions of the northern United States, on the high mountain plateaus, and between the Sierra Madre Oriental and the Sierra Madre Occidental in Mexico. Red Desert and Desert soils are extensive in the hotter arid regions from Coahuila and Texas to the west coast.

Summer temperatures are high everywhere that Aridisols and associated soils are mapped in the United States and Mexico. The soil temperature regime may be mesic, thermic, or hyperthermic. The soil moisture regime is aridic, torric, or ustic.

AR1. Gypsiorthids — Calciorthids — Paleorthids

This soil association occupies the Pecos River Valley of New Mexico and Texas and the adjoining uplands. Gypsum and limestone make up most of the surficial geologic material in the area. There are thick deposits of sylvite, a muriate of potash ore, near Carlsbad (New Mexico) which represent the major U.S. source of potash. Groundwater is plentiful but much of it is saline, particularly in the southern part of the area, as is the Pecos River. The salts are mainly sodium and potassium chlorides and sulfates, calcium bicarbonate, and calcium sulfate. The first detailed study made in the U.S. of the relation between soil salinity and crop growth was conducted by C.S. Schofield in the Pecos Valley in 1939 and 1940 (Schofield, 1941). Native vegetation in the Pecos Valley consists of grasses (*Bouteloua, Agropyron, Muhlenbergia, Andropogon, Sporobolus*) and shrubs (*Larrea, Prosopis, Atriplex, Flourensia*). The topography is broad gently undulating plains and rolling low hills. Annual precipitation amounts to 300—400 mm and the growing season exceeds 200 days.

Gypsiorthids have a gypsic (sometimes petrogypsic) horizon at 10—50 cm and usually have a calcic horizon as well. Surface soils are pale-brown to light brownish-gray loams; subsoils are brown or brownish-gray loams. The soils are calcareous throughout the profile and become strongly calcareous and saline in the gypsic horizon. Permeability is moderate. Organic-matter content is low and soil reaction is moderately alkaline.

Calciorthids have a calcic horizon at 10—30 cm, generally, and may be underlain by beds of gypsum in this area. Color of the surface soil is brown and the subsoil is brownish gray to light gray. Loam textures are common for the surface soil and the subsoil. Most profiles are calcareous in the surface as well as in the subsoil and many of the soils are slightly to moderately saline. Soil reaction is slightly to moderately alkaline. Organic-matter content is low and permeability is good.

Paleorthids are Aridisols with a petrocalcic (hard caliche) horizon at 10—30 cm which is overlain by a brown calcareous loam. The upper centimeter of the petrocalcic horizon usually consists of very hard laminated calcium carbonate, whereas the underlying carbonate is less hard. Gypsum is intermingled with calcium carbonate in the lower part of the petrocalcic horizon. Permeability to water depends upon the amount of cracking of the petrocalcic horizon. Cracks are common, so water seldom is perched above the indurated carbonate layer for many hours.

Other great groups in this association are Camborthids, Salorthids, Torripsamments, Torriorthents, and Torrifluvents.

AR2. Calciorthids — Paleorthids — Orthents

Soils of this association are very shallow to moderately deep, calcareous, frequently stony or gravelly, and occur on landscapes combining broad rolling plains, alluvial fans, river valleys, mesas, and hills. Little or no development is apparent in most of the soils. Vegetative cover consists of grasses and scattered shrubs and trees. Among the grasses, *Hilaria*, *Andropogon*, *Agropyron*, and *Stipa* are dominant in the north, and *Bouteloua*, *Muhlenbergia*, *Hilaria*, and *Sporobolus* in the south. The principal shrubs in the north are *Artemisia*, *Chrysothammus*, and *Atriplex*, whereas in the south they are *Larrea*, *Prosopis*, and *Flourensia*. Pinon (*Pinus* spp.) and juniper (*Juniperus* spp.) are found fairly commonly. Annual precipitation averages between 150 and 400 mm, more than half of it coming during the warm season. The growing season is short in the north (90—130 days) and longer in the south (150—230 days).

Calciorthids have an A horizon which is 10—20 cm thick, is pale brown to dark brown in color, has a loam or sandy loam texture, is calcareous, has a low to moderate organic-matter content, and is slightly to moderately alkaline. There is no B horizon but there is a calcic horizon immediately underlying the A horizon. The calcic horizon often is gypsic in its lower part and frequently contains hard nodules of calcium carbonate. Calciorthids are prevalent on the rolling plains. Some of them are covered by a desert pavement.

Paleorthids are Calciorthids with a petrocalcic horizon instead of a calcic horizon. Generally, there are numerous cracks in the indurated calcium-carbonate horizon which permit roots and water to penetrate into the usually softer carbonate material below. The silica content of the petrocalcic horizon sometimes is considerable, indicating that cementation is due to precipitation and dehydration of both carbonates and silica. Petrocalcic horizons which are only moderately hard upon initial exposure in road cuts commonly become very hard after several months of further dehydration. Paleorthids occur on the more level upland sites. Desert pavements are common.

Orthents include Torriorthents and Ustorthents. They are coarse textured, usually gravelly and frequently stony, and of variable depth over bedrock. On alluvial and colluvial fans, soil depth varies from shallow to deep whereas on steep slopes and on the tops of hills and mesas, the soils are shallow or very shallow. While the surface soil may be noncalcareous, subsoils usually are calcareous. Organic-matter content ranges from very low to moderately high. Soil color may be reddish brown, brown, or brownish gray. Most of the surface soils are covered with a desert pavement.

Among the other great groups are Pellusterts, Torrifluvents, Ustifluvents, Salorthids, Natrargids, and Torripsamments.

AR3. Calciorthids — Orthents — Haplargids

This combination of soils has the greatest areal extent of any association, going from central Mexico to the northwestern United States, a distance of more than 3,000 km. The landscape in Mexico, New Mexico, Idaho, and Washington is that of a broad undulating plain cut by numerous drainageways and broken by hills and low mountains. In the Great Basin centered on Nevada, the landscape is a repeating sequence of mountains, fans, and basins. Elevations go from sea level to 3,000 m but most of the area is below 1,500 m. Precipitation is as low as 40 mm in eastern California and as high as 400 mm or more in San Luis Potosí. Growing seasons vary in length from 130 days on the Colorado Plateau to 365 days on the Gulf of California.

Calciorthids are the dominant soils on the broad undulating plains between the mountains and the valleys and basins. The A horizon is light brown to grayish brown in color, 5—15 cm thick, low to moderate in organic matter, calcareous, moderately alkaline, sandy loam to clay loam in texture, and commonly has a pebbly desert-pavement cover although a thin cover of loose aeolian sand sometimes is present. Usually, the upper part of the B horizon is transitional to the calcic horizon which begins at 20—35 cm below the soil surface. The calcic horizon is generally one or more meters in thickness. Subsoils are sandy loams to clay loams and may be gravelly; they frequently are brown or reddish brown in the upper part and white or pinkish white in the lower part. Hard calcium carbonate nodules are present in the lower part of most Calciorthid profiles and gypsum may be present.

Orthents consist of Torriorthents in the drier areas and Ustorthents in the moister areas. They occur on the colluvial and alluvial fans and on steep mountain slopes. Fan soils are shallow to deep; mountain soils are shallow or very shallow. Most of the soils are calcareous to the surface; low in organic matter; brown, grayish-brown, or reddish-brown in color; and have gravelly or stony sandy loam textures.

Haplargids are the developed soils of the arid regions. Typically, the A horizon is reddish brown, 5—20 cm thick, low in organic matter, non-calcareous or slightly calcareous, sandy loam to loam in texture, platy and massive or weak granular in structure, slightly acid to slightly alkaline in reaction, and has a desert pavement of angular stones. The well-defined argillic B horizon is as red or redder than the A horizon, 30 to more than 100 cm thick, has prismatic or blocky structure, is moderately to strongly alkaline, and is calcareous. Carbonate nodules may be present in the lower part of the B horizon.

Camborthids, Paleorthids, Torrerts, Torripsamments, Torrifluvents, and Ustifluvents are other soils commonly found in this association.

Entisols

The Entisol order in arid North America includes a variety of soils: shallow to deep stratified recent alluvium, shallow stony soils over bedrock, deep gravelly soils on alluvial fans, sand dunes, and others. The one thing they have in common is that they show little or no evidence of soil development. Most of the Entisols are used for grazing but the Fluvents may be highly productive soils when irrigated.

Soil moisture regimes are aridic, torric, or ustic, mainly, and sometimes udic. Soil temperature regimes range from mesic to hyperthermic.

E1. Torriorthents — Haplargids

Southeastern Colorado is an undulating to hilly grassland region cut by the Arkansas River. Generally speaking, Haplargids with inclusions of Torriorthents, Camborthids, and Torripsamments are dominant north of the river and Torriorthents with inclusions of Haplargids and Argiustolls are dominant south of the river. The principal grasses are *Agropyron*, *Festuca*, *Poa*, *Bouteloua*, and *Hilaria*. Sagebrush (*Artemisia* spp.) is the main shrub. Annual precipitation is about 350—400 mm, most of which falls from April to September. Temperature ranges are great, with maxima around 45°C, minima around —32°C, and the annual mean about 12°C. The growing season lasts for about 170 days.

Torriorthents are shallow or very shallow loam-textured soils over bedrock. Many are stony, low to moderate in organic-matter content, brown in color, noncalcareous in the surface, and calcareous just above bedrock. They occupy the crests and upper slopes of hills.

Haplargids are moderately deep to deep Aridisols with a thin (15—20 cm) surface horizon and an argillic B horizon. Surface-soil texture varies from sandy loam to silt loam and subsoil texture from loam to silty clay loam. Soil colors are some shade of brown, usually dark brown in the surface and yellowish brown in the subsoil. Soil reaction ranges from moderately acid in the surface to moderately alkaline in the subsoil. There usually is an accumulation of calcium carbonate beginning in the lower part of the B horizon or in the C horizon. Organic-matter content is less than 1.5% in the surface.

Other important soils in this association are Camborthids, Torripsamments, Argiustolls, and Ustifluvents.

E2. Ustipsamments

The Nebraska Sand Hills consists of a succession of sand-dune hills and ridges in a rolling topography having occasional intervening valleys and depression. There are very few surface streams because rainfall is quickly absorbed by the porous sand. The aeolian sand is derived from Tertiary sandstone and most of the dunes are covered with a grass sod. Grasses are dominant. They include *Andropogon*, *Bouteloua*, *Sporobolus*, and *Calamovilfa*. The main shrubs are *Artemisia* and *Yucca*. Precipitation averages between 400 and 600 mm, the mean annual temperature is about 8°C, the growing season is about 140 days, and wind velocity is high.

Ustipsamments in the Nebraska Sand Hills are deep, have a dense grass cover and a noticeably darker A horizon a few centimeters thick overlying a lighter grayish-brown fine and medium sand. The soils are noncalcareous and slightly to moderately acid. They erode badly when the grass cover is disturbed.

The other significant great group is the Ustorthents.

E3. Torriorthents — Haplargids — Natrargids

The soils of this association occupy a treeless arid and semiarid region on the western part of the Great Plains which has in it a humid island consisting of the Black Hills. Topographically the region is a succession of smooth watershed divides with intervening valleys of rolling or rough land. A goodly part is made up of barren excessively eroded badlands east and north of the Black Hills showing little soil development. Shale is the source of sodium salts which have resulted in the formation of Natrargids. Extensive coal deposits underlie most of the region. Average annual precipitation is about 250 mm in the west and 400 mm in the east. The annual temperature is about 7°C, summers are hot, winters are cold, and the growing season is between 100 and 140 days long. Grasses such as *Bouteloua*, *Buchloe*, and *Agropyron* are dominant; sagebrush (*Artemisia*) and shadscale (*Atriplex confertifolia*) are the principal shrubs but trees of *Pinus* species also are present.

Torriorthents are shallow and very shallow soils overlying shale, sandstone, and limestone. They are concentrated in Montana and South Dakota. Very shallow soils preveail on the shale sites; shallow soils are more common on sandstone and limestone. Slopes are steep in the shale and sandstone badlands but range from nearly level to steep in the other Torriorthent areas. Erosion is excessive on the shale beds, canyons are deep, and valley bottoms are narrow. The soils may be calcareous to the surface and calcium carbonate usually is abundant above the bedrock. Organic-matter content is low to moderate, textures vary from clay loam to sandy loam, and soil depth averages less than 20 cm over bedrock. The shale Torriorthents may be saline and highly variable in color.

Haplargids have thin A horizons over the argillic B horizon and are found on gently undulating landscapes. Surface soils are 10—15 cm thick, dark brown or dark grayish brown in color, noncalcareous, slightly alkaline, medium to fine textured, and low in organic matter. Subsoils are finer textured than the surface soils, have a yellowish-brown color and a blocky or prismatic structure, and usually are calcareous in the lower part.

Natrargids occur on level or gently undulating landscapes and are intermingled with Haplargids. Many of them are saline in or below the natric horizon. Surface soils are about 10 cm thick, brownish gray in color, silt loam to clay loam in texture, low in organic matter, and frequently have a layer of dispersed soil on the surface which resists water penetration. Soils exhibiting the latter condition are called "slick spot" soils. Subsoils are finer textured than the surface soils, have a prismatic or columnar structure, and are slowly or very slowly permeable. The exchangeable sodium percentage of the natric horizon is, by definition, greater than 15. Soil reaction is moderately alkaline in the surface and strongly alkaline in the natric horizon. The natric horizon usually is 25—50 cm thick and contains in its lower part a calcium carbonate horizon which frequently is gypsiferous and may be saline.

There are inclusions of Camborthids, Haplustolls, and Ustifluvents among the great groups in this association.

E4. Torrifluvents — Salorthids

The Imperial and Mexicali valleys, with their extension into the Coachella Valley, the Yuma area, and the flood plain at the mouth of the Colorado River, represent the largest body of Fluvents in arid North America. Between the Imperial and Coachella valleys is the Salton Sea which lies below sea level. The Salton Sea was formed when the Colorado River accidentally was diverted into the Salton Depression for 18 months during the period from 1905 to 1907.

More than 400,000 hectares of land are irrigated by Colorado River water in this part of the United States and Mexico. Salinity problems have become acute in the area, despite the installation of a comprehensive drainage system in the Imperial and Coachella valleys, due to rising salinity of the river water. Cotton, alfalfa, barley, and vegetables are the principal crops. Temperatures are high all the year around, with an annual average of about 22°C, and the growing season is close to 365 days. Precipitation amounts to only 60—

90 mm. Vegetation consists of shrubs and trees such as *Prosopis* and *Tamarix* and of *Bouteloua* and *Hilaria* grasses. The topography is level.

Torrifluvents are deep, generally medium to fine textured in the surface and in the subsoil, stratified, variable over short horizontal distances, low in organic matter, and moderately alkaline in reaction. Most of the Torrifluvents are calcareous throughout the profile, are moderately to slowly permeable, and contain moderate amounts of soluble salts.

Salorthids are Torrifluvents in which a salic horizon has developed as the result of a high water table. In most cases, the salic horizon begins at the soil surface. In other profile characteristics, Salorthids resemble Torrifluvents except that they have a water table within 1 m of the surface for much or all of each year. Salorthids are commoner in the Mexicali Valley than in the Imperial Valley which adjoins it and is in the same geomorphic unit.

Torriorthents and Torripsamments also are found in this association, especially along the east side.

E5. Torriorthents — Xerochrepts — Durixeralfs

The Central Valley of California has a northern part called the Sacramento Valley and a southern part known as the San Joaquin Valley. The two meet at the peat lands of the Sacramento—San Joaquin Delta east of San Francisco.

An east—west cross-section of the Central Valley would show a river and its flood plain in the center, flat basins on either side of the river, gently sloping alluvial fans above the basins and extending to terrace deposits, and rolling foothills and steep mountains above the terraces. Alfisols in varying stages of development are found in the rolling foothills, on the terrace deposits, and on the alluvial fans. Entisols are intermingled with the Alfisols but also are the major flood-plain soils. Inceptisols and Aridisols are confined largely to the alluvial fans. Mollisols are important among the basin soils and Histosols make up the peat and muck soils.

Summers are hot and dry, winters are cool and relatively moist. Rainfall increases from south to north, with an average of about 150 mm per year in the south part of the valley and approximately 500 mm in the extreme north. Annual temperatures reach their lowest at 15°C in the Delta and are highest at the north and south ends of the Central Valley where the average is 17°C or 18°C. Growing seasons are 240—300 days long.

Torriorthents are undeveloped alluvial fan soils, brownish gray in color, calcareous to the surface, medium textured, and low in organic matter. Gypsum crystals sometimes are present below 40 cm. They are moderately deep to deep over rock and occupy gentle slopes.

Xerochrepts also are alluvial fan soils but they differ from the Torriorthents in that they show a slight degree of development in the form of a cambic horizon and occur where the precipitation exceeds 350 mm. They are formed on recent alluvium of basic igneous or sedimentary rock origin. Surface soils are brown in color, weakly granular, coarse to fine textured, low in organic matter, neutral in reaction, and about 25—30 cm thick. Subsoils are lighter in color, blocky in structure, somewhat finer textured than the surface soil, neutral in reaction, and 20—40 cm thick. Bedrock underlies the soils at 60—120 cm, usually.

Durixeralfs are the Alfisols with, in this case, an iron and silica-cemented hardpan (duripan) at 50—80 cm and about 20 cm thick. The argillic horizon begins at 15—20 cm above the duripan. These soils occur on old terraces that are very gently sloping or undulating and have a hummocky microrelief. The parent material is moderately coarse-textured alluvium of dominantly granitic origin. Permeability of the duripan horizon is slow. Brown is the customary surface-soil color and brownish red or gray the usual subsoil color. The typical surface-soil texture is fine sandy loam or loam and the texture of the argillic horizon is sandy clay loam or clay loam. Many of the Durixeralfs are calcareous in the subsoil and some are calcareous in the surface soil, as well. Durixeralfs having a natric horizon and containing large amounts of soluble salts are extensive in the San Joaquin Valley.

There are several other great groups of importance in the Central Valley. Among them are Haploxerolls and Haplaquolls in the fine-textured basin soils, Xerofluvents in the river flood plains, Xeropsamments in the southern San Joaquin Valley, and Haploxeralfs and Palexeralfs in the north and east.

Mollisols

The dominant soils throughout the length and breadth of the Great Plains are Mollisols. While Great Plains Mollisols are mostly grassland soils, the Mollisols in Mexico and the western half of the arid United States usually are woodland and shrubland soils having grasses mixed in with the woody vegetation. Mollisols have a friable mollic epipedon which is a relatively thick dark-colored mineral surface horizon rich in organic matter, the cation-exchange complex is dominated by calcium and magnesium, and the base saturation is high. In arid North America, the soil temperature regime is frigid, mesic, thermic, or hyperthermic and the soil moisture regime is ustic or xeric. Montmorillonitic clay minerals are dominant. Many of the Mollisols have a calcic horizon and an argillic horizon, and a few have a natric horizon. Climatically, the eastern part of the Mollisol associations in the Great Plains lies in subhumid rather than semiarid regions.

M1. Calciustolls — Haplustolls

Mollisols with a thin A horizon and calcareous profiles are the principal soils over a large part of central Texas and northeastern Mexico. The topography is variable, from an undulating plain west of the Rio Bravo (Rio Grande) in Mexico to plateaus, hills, and low mountains in the remainder of the region. Annual precipitation varies from 450 to 600 mm, most of it coming in spring and summer. Annual temperatures are between 18 and 22°C and the length of the growing season is from 210 days in the north to more than 300 days in the south. The original vegetation was a grassland of *Bouteloua*, *Andropogon*, *Panicum*, and *Buchloe* species but shrubs and trees of *Prosopis*, *Quercus*, and *Juniperus* now dominate the vegetative cover.

Calciustolls are shallow to moderately deep over limestone in this area. Most of them are shallow over a petrocalcic (cemented calcium carbonate) horizon. Organic-matter content is moderate. Typically, the Petrocalcic Calciustoll has an A horizon which is a grayish-brown loam, or clay loam, about 12 cm thick, containing numerous fragments of indurated caliche or limestone. Rock or caliche fragments cover a considerable amount of the surface. Beneath the A horizon is a 5—15 cm thick C_1 horizon of cemented caliche fragments which represent about 90% of the soil mass. The C_2 horizon consists of massive calcium carbonate accumulations more than 1 m thick containing indurated caliche nodules. Typic Calciustolls are slightly calcareous in the surface, have a calcic horizon at about 20 cm, and usually are underlain with limestone at depths between 40 and 80 cm. The Calciustoll topography is nearly level to gently sloping.

Haplustolls formed in undulating to hilly areas and are very shallow over hard fractured limestone. Textures range from loam to clay, organic-matter levels are moderate, and surfaces are about half covered with boulders, stones, and pebbles of limestone. Surface soils are dark brown, calcareous, and pebbly. Limestone underlies the surface soil at about 20 cm.

Other great groups of significance in this association include Ustorthents, Pellusterts, Calciorthids, and Paleorthids.

M2. Calciustolls — Ustochrepts — Paleustalfs

The topography of this association ranges from nearly level to gently sloping over most of the area but in places there are steep escarpments and rough land dissected by numerous drainage channels. While most of the soils are moderately deep and deep, there are many shallow and very shallow soils over limestone, shale, and gypsum. Precipitation increases from west to east, from a low of about 450 mm to a high of 650 mm. Most of the precipitation comes during the April—October period. Summers are hot and winters are cold. Annual temperature in the west is approximately 13°C and in the east is 17°C. Growing seasons are a little more than 200 days. The area is a grassland of *Bouteloua*, *Andropogon*, *Buchloe*, and *Agropyron* and other genera. *Prosopis* shrubs are invaders when the land is overgrazed.

Calciustolls are calcareous loamy soils occupying gentle slopes. The surface horizon has a loam to clay loam texture, is brown in color, has a moderate level of organic matter, is calcareous, and contains a few calcium carbonate nodules. In this association, most Calciustolls have a cambic B horizon. The calcic horizon may begin with the cambic horizon or deeper in the profile. The C horizon appears at depths varying from 25 to 80 cm.

Ustochrepts are weakly developed soils derived from Permian red bed calcareous clayey shales, from sandy alluvium, and from sandy aeolian deposits. Those derived from shales are the most widespread. The latter have brown to reddish-brown clay loam surface soils 15—20 cm thick overlying a reddish-brown clay loam or clay cambic B horizon 25 cm thick which, in turn, overlies clayey shale. They are calcareous to the surface, are low in organic matter, and may or may not have a calcic horizon. The topography is gently sloping to steep uplands. The sandy Ustochrepts have a brown to grayish-brown fine sandy loam surface soil which is 12 to 50 cm thick and overlies a reddish-brown to brown cambic B horizon about 50 cm thick. Subsoils of the sandy Ustochrepts are calcareous but the surface soil is either noncalcareous or weakly calcareous.

Paleustalfs are represented by deep sandy to loamy soils on nearly level to sloping relief of the uplands. Many of them have hummocks of sand on the surface. The argillic B horizon begins at depths of 25—50 cm and is 1 m or more in thickness. The lower part of the B horizon and the C horizon may be calcareous. Surface soils are brown, fine sand to fine sandy loam in texture, low in organic matter and neutral to slightly alkaline in reaction. Subsoils are reddish brown, have a fine sandy loam to sandy clay loam texture, and are neutral to slightly alkaline in the upper part. Wind erosion is a major hazard on these soils. Paleustolls and Argiustolls are other important soils. There also are significant amounts of Ustipsamments, Calciorthids, and Torriorthents in this association.

M3. Argiustolls — Paleustolls — Haplustolls

This association of soils goes from Texas to South Dakota, encompasses most of the middle part of the arid Great Plains, and extends into the subhumid Great Plains. The topography is level to undulating and the dominant textures are silt loam and clay loam. The entire area is a grassland of *Bouteloua*, *Agropyron*, *Buchloe*, and *Stipa* genera. The sparse shrub population is represented mainly by *Artemisia* in the north and *Prosopis* and *Yucca* in the south. Annual precipitation increases from west to east; annual temperature generally increases in the same direction but there also is a marked north—south temperature increase. Mean precipitation ranges from 350 to 550 mm and mean temperature from 8 to 17°C. Growing seasons last for 140—190 days.

Argiustolls are called hardland soils because surface soils have a silt loam or clay loam texture and the subsoils are a moderately compact clay loam or clay. They occur on the nearly level uplands of the plains. Color of the surface soil and the subsoil is dark grayish brown or dark brown. Organic-matter content is between 1.5 and 2% for the surface soil, which is from 10 to 30 cm thick. The argillic horizon is 60—150 cm thick and may be calcareous in its lower part. Calcic horizons begin at 30—75 cm. Surface soils are slightly acid and subsoils are neutral to moderately alkaline. Argiustolls are among the best soils for producing dryland wheat and are good soils for irrigated crops. They are moderately or slowly permeable.

Paleustolls also are deep hardland soils with profile characteristics similar to those of the Argiustolls, but with a more compact and clayey argillic subsoil. The argillic horizon usually is more than 1 m thick. Subsoil colors are more reddish than in the Argiustolls. The calcic horizon begins at 75—150 cm. Soil reaction above the calcic horizon is slightly alkaline. Permeability is slow to very slow.

Haplustolls are medium to fine textured deep soils on nearly level to gently sloping uplands. They have a weakly developed cambic and calcareous horizon and a calcic horizon which may begin in either the B or C horizon. They sometimes are calcareous to the surface and are slightly to moderately alkaline. Surface soils are grayish brown and subsoils are grayish brown or yellowish brown. Permeability is moderate.

Among the other great groups represented in this association are Torriorthents, Ustorthents, Ustipsamments, Ustifluvents, and Natrustolls.

M4. Argiborolls — Ustorthents — Natriborolls

The plains of Alberta, Saskatchewan, Montana, North Dakota, and South Dakota are the locale for this association. The western border is the Rocky Mountains, from which the plains slope gently to the east. Annual temperatures range from 2 to 8°C and temperature extremes are pronounced. At Bismarck, North Dakota, the record maximum was 46°C and the record minimum was —43°C, for a difference of 89°C. Most weather stations show differences of at least 80°C between the maximum and minimum record temperatures. The driest part of the region has an annual precipitation of about 300 mm and the wettest part has about 450 mm. Summer is the wet season and winter is the dry period. Growing seasons amount to 100—140 days. The soil temperature regime is frigid. Vegetation consists of grasses such as *Agropyron*, *Bouteloua*, *Andropogon*, and *Elymus*.

Argiborolls have a dark-grayish brown medium to fine textured surface soil underlain at 20—25 cm by a finer-textured argillic horizon having a brown color. Most Argiborolls are calcareous in the lower part of the argillic horizon. Organic-matter content is moderate and soil reaction is neutral to slightly alkaline in the surface. The topography is level to gently undulating and is broken by valleys with gentle to steep slopes. Most of the soils are moderately deep and slowly permeable.

Ustorthents are the shallow soils of hilltops and steeper slopes of the valleys. Textures are loams and clay loams, organic-matter content is low, soil reaction is neutral to slightly alkaline, a calcareous layer may occur just above the bedrock, and the profile frequently is stony.

Natriborolls have a natric horizon at a depth of about 15—20 cm, usually are moderately saline, and have an accumulation of calcium carbonate and threads of gypsum in the lower part of the natric horizon. Surface soils are brown and brownish-gray loams and silt loams and subsoils are dark-brown silty clay loams or clay loams. The natric horizon is slowly or very slowly permeable. Many of the Natriborolls have an albic (bleached) horizon above the natric horizon and those known as solodized solonetz have an acid surface soil. Natriborolls are most likely to be found in slight depressions on the level to gently undulating landscape. They are extensive in Alberta and Saskatchewan and are less common in Montana and North Dakota.

Torriorthents, Ustifluvents, and Haploborolls are other great groups of importance in this association.

M5. Haplustolls — Calciustolls — Torriorthents

Mollisolic soils are common constituents of the soil complex in the mountain foothills, plateaus, and mesas west and south of the Great Plains. The topography differs from that of the plains in that it ranges from gently undulating to sloping. Likewise, the vegetation consists of a shrub-grass type rather than of a grassland. *Artemisia*, *Chrysothamnus*, *Quercus*, *Pinus*, and *Juniperus* are the dominant shrubs and trees. The native grasses are mainly *Agropyron*, *Festuca*, *Bouteloua*, and *Poa*. Precipitation and temperature are lower in the Columbia Plateau and along the west side of the Rocky Mountains than they are in the foothills of the Sierra Madre Occidental. The range in annual precipitation is from about 200 to 500 mm and the temperature range is from 8 to 13°C. The growing season may be as short as 120 days in southeastern Oregon; it probably exceeds 200 days in central Mexico.

Haplustolls have a prominent surface horizon with dark-brown to grayish-brown colors, a moderate amount of organic matter, and a neutral to slightly alkaline reaction. Surface-soil textures are fine sandy loams to silt loam. The B horizon, which begins at depths of 10—40 cm, is alkaline and commonly has more clay than the A horizon but not enough to qualify as an argillic horizon. Subsoil structure is developed sufficiently to produce a cambic horizon. Typically, calcium carbonate has accumulated in the lower part of the B horizon and the upper part of the C horizon. The topography in the Columbia Plateau is mostly undulating, whereas it is steeper in the remainder of the region.

Calciustolls are calcareous to the surface and have a calcic horizon at about 25—40 cm. The calcic horizon normally exceeds 20 cm in thickness and may be several meters thick. Organic-matter content is moderately high. Surface-soil color is pale brown to grayish brown and subsoil color is yellowish brown. Textures range from sandy loam to clay loam. Most of the soils are moderately deep but shallow soils are common, especially in Mexico. Petrocalcic horizons sometimes are present at depths of 60—100 cm. Calciustolls generally are found on nearly level and gently sloping surfaces.

Torriorthents are mostly shallow and very shallow soils over bedrock, but some of them are moderately deep to deep gravelly alluvial fan soils. Textures are variable and typically the shallower Torriorthents are stony. Organic-matter content is low, soil reaction is slightly alkaline, and there may be calcium carbonate accumulations just above bedrock in the shallow soils. Slopes are moderate to steep. Among the other great groups are Paleustolls, Arguistolls, Calciorthids, and Paleorthids.

Vertisols

Deep cracking clays derived from marine and deltaic calcareous deposits are extensive in the Rio Grande Plain and along the coast of the Gulf of Mexico. Vertisols occur in other parts of arid North America but they do not begin to compare in area with those in the Texas—northeastern Mexico region. During dry periods, Vertisols develop cracks which are several centimeters across at the surface and which extend downward for more than 50 cm. When wet, the montmorillonitic clay minerals expand to close the cracks. Pressures generated when the wet soil expands cause an upward movement of soil which results in formation of the microrelief known as gilgai. The soil movement can break highway pavement and masonry building foundations and tilt trees and power line poles. Due to their very low permeability, Vertisols show little or no evidence of development. The soil moisture regime is ustic and the soil temperature regime is hyperthermic. About 80% of the soils are Vertisols, 10% are Paleorthids, and the remainder are a mixture of several great groups.

The topography is level to gently undulating throughout most of the region. Annual precipitation varies from 400 to 700 mm, being higher near the coast and toward the south. Floods and droughts are common, with both sometimes occurring in the same year. Annual temperatures are between 21 and 23°C. Growing seasons are about 330 days along the coast and 310 days inland. The native vegetation was grassland with some shrubs but shrubs have increased greatly due to grazing pressures. The dominant grasses are *Bothriochloa*, *Digitaria*, *Buchloe*, and *Hilaria*. The shrubs and trees are *Quercus*, *Prosopis*, *Cactaceae*, and *Acacia*.

V. Pellusterts — Chromusterts — Paleorthids

Pellusterts are the Vertisols with gray to black colors in the upper 30 cm and having cracks which open and close once or twice each year, depending upon how wet or dry a particular year may be. They are clay loams and clays and are calcareous throughout the profile. Organic-matter content is low and permeability is very slow. Water penetration into the subsoil would be minimal if it were not for all the deep cracks into which water runs. The soils are deep and lie on level or depression surfaces where external drainage is restricted and water stands on the surface.

Chromusterts are identical to Pellusterts in morphological characteristics but the soil color is brownish or reddish rather than gray or black. The surface usually is darker than the subsoil. Internal drainage is very slow, but external drainage is better than with the Pellusterts because Chromusterts occur mostly on gentle slopes where water does not stand. They are deep and calcareous.

Paleorthids have a petrocalcic horizon at 20—25 cm which is strongly cemented at the top but becomes less cemented with depth. Surface soils are usually grayish-brown to dark-brown calcareous gravelly or stony loams which overlie abruptly the petrocalcic horizon. The soils occupy undulating to hilly ridges.

Among the numerous great groups in this association are Torrerts, Ustochrepts, Salorthids, Ochraqualfs, Argiustolls, Haplustolls, and Udipsamments.

REFERENCES

Aandahl, A.R., 1972. *Soils of the Great Plains.* Lincoln, Nebraska, scale 1:2,500,000.

FAO/UNESCO, 1972. *Soil Map of the World, 1:5,000,000. North America.* UNESCO, Paris, two sheets.

Meigs, P., 1953. World distribution of arid and semi-arid homoclimates. In: UNESCO, *Reviews of Research on Arid Zone Hydrology. Arid Zone Res., I:* 203—210.

Schofield, C.S., 1941. *The Pecos River Joint Investigation*, 1939—1940. *Soil Salinity Investigation.* Bureau of Plant Industry, U.S. Department of Agriculture, Washington, D.C., 191 pp. (mimeograph).

Secretaría de Recursos Hidráulicos, 1970. Unidades de Suelos de la República Méxicana segun el Sistema de Clasificación de FAO/UNESCO. Primer Intento. Escala 1:1.000.000. In: *Informe de Actividades de la Dirección de Agrología, Mayo 1967—Septiembre 1970.* Publicación Numero 1. Dirección de Agrología, Jefatura de Irrigación y Control de Ríos, Secretaría de Recursos Hidráulicos, México, D.F.

U.S. Department of Agriculture, 1941. *Climate and Man. Yearbook of Agriculture, 1941.* U.S. Government Printing Office, Washington, D.C., 1248 pp.

U.S. Department of Agriculture, 1964. *Soils of the Western United States.* Soil Conservation Service, U.S. Department of Agriculture, Washington, D.C., 69 pp.

U.S. Department of Commerce, 1969. *Climates of the World.* Environmental Data Service, Environmental Science Services Administration, U.S. Department of Commerce, Washington, D.C., 28 pp.

Chapter 8

SOUTH AMERICA

INTRODUCTION

Most of the arid regions of South America are concentrated in the southern part of the continent and along the west coast, extending 5,500 km from the cool deserts of Patagonia to the extremely arid coastal deserts of Peru (Fig.8.1). A large area in northeastern Brazil and a smaller one in northern Colombia and Venezuela also are arid. One of the two driest places on the face of the earth is the Atacama Desert in northern Chile (the other is the central Sahara) where the average annual precipitation may be less than 1 mm. Several rivers heading in the Andes cross the arid part of Argentina and Paraguay and 52 small rivers flow across the narrow Peruvian coastal plain. There is one major river, the Sao Francisco, in northeastern Brazil.

Alfisols, Aridisols, Entisols, and Mollisols are the principal soil orders in arid South America (Fig.8.2). The main sources of soil information for the continent were Volume IV, South America, of the FAO/UNESCO Soil Map of the World project (1971), a paper by Papadakis (1963), and a map by Etchevehere (1971).

ENVIRONMENT

Climate

Annual precipitation in the South American arid region varies from nearly zero in the Atacama Desert to over 700 mm in northeastern Brazil. Annual temperatures show much less variation, the coldest places being in Patagonia and on the Altiplano of the Andes Mountains and the hottest in the tropical north.

Data on precipitation and temperature for five locations in South America are given in Table 8.1. Rainfall distribution patterns differ considerably at the various stations. Maracaibo, in the north, has a moist winter, a wet spring, and a dry summer; in the northeast, Quixeramobim, representative of northeastern Brazil uplands, has a wet autumn but is dry for the remainder of the year; Lima, on the west coast of Peru, has little rain at any time; Mendoza, located in the shadow of the Andes, receives most of its modest amount of precipitation in the winter; and Bahia Blanca, on the east coast of Argentina, has only moderate variation from season to season. At 752 mm,

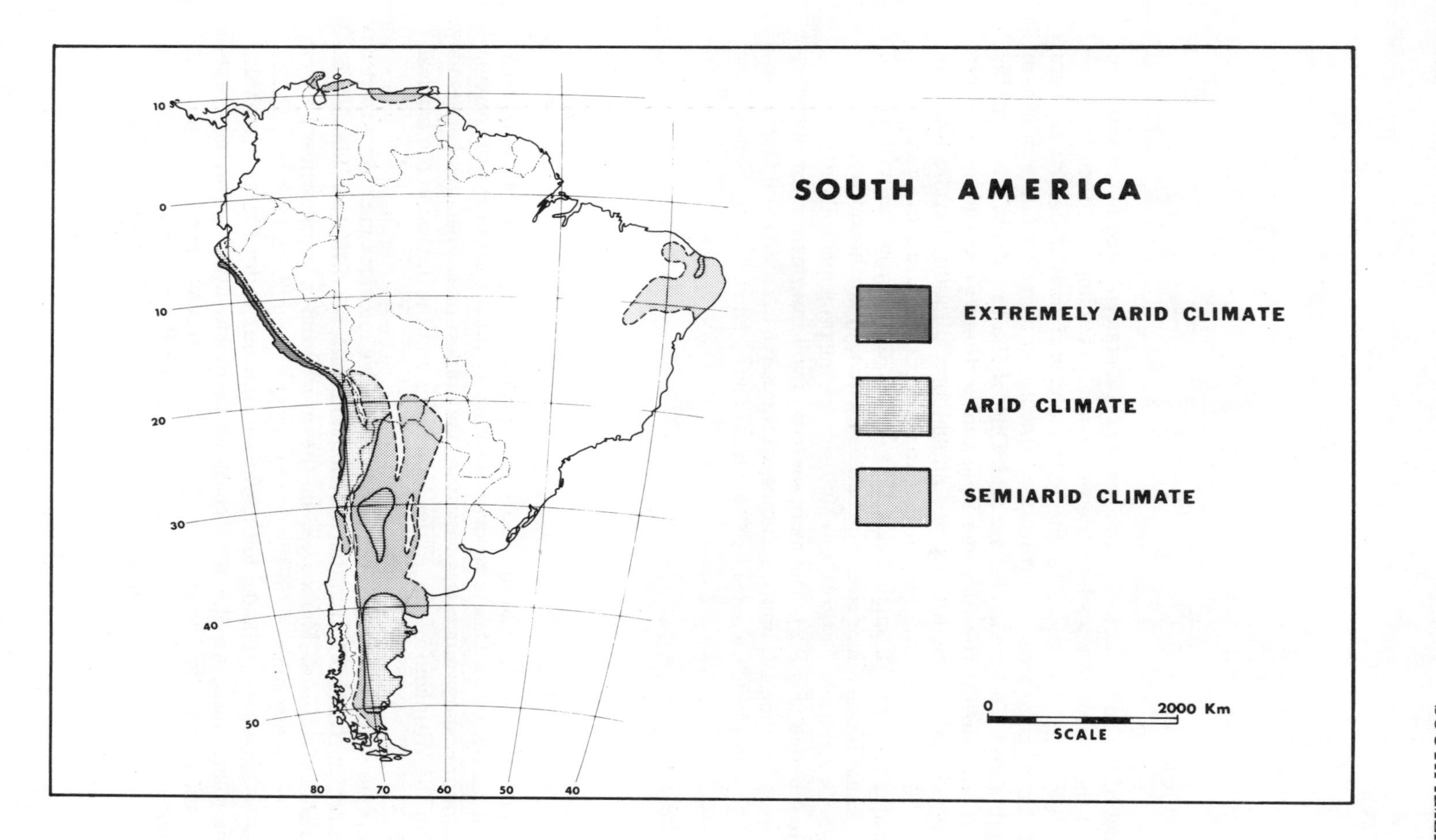

Fig.8.1. Arid regions of South America (after Meigs, 1953).

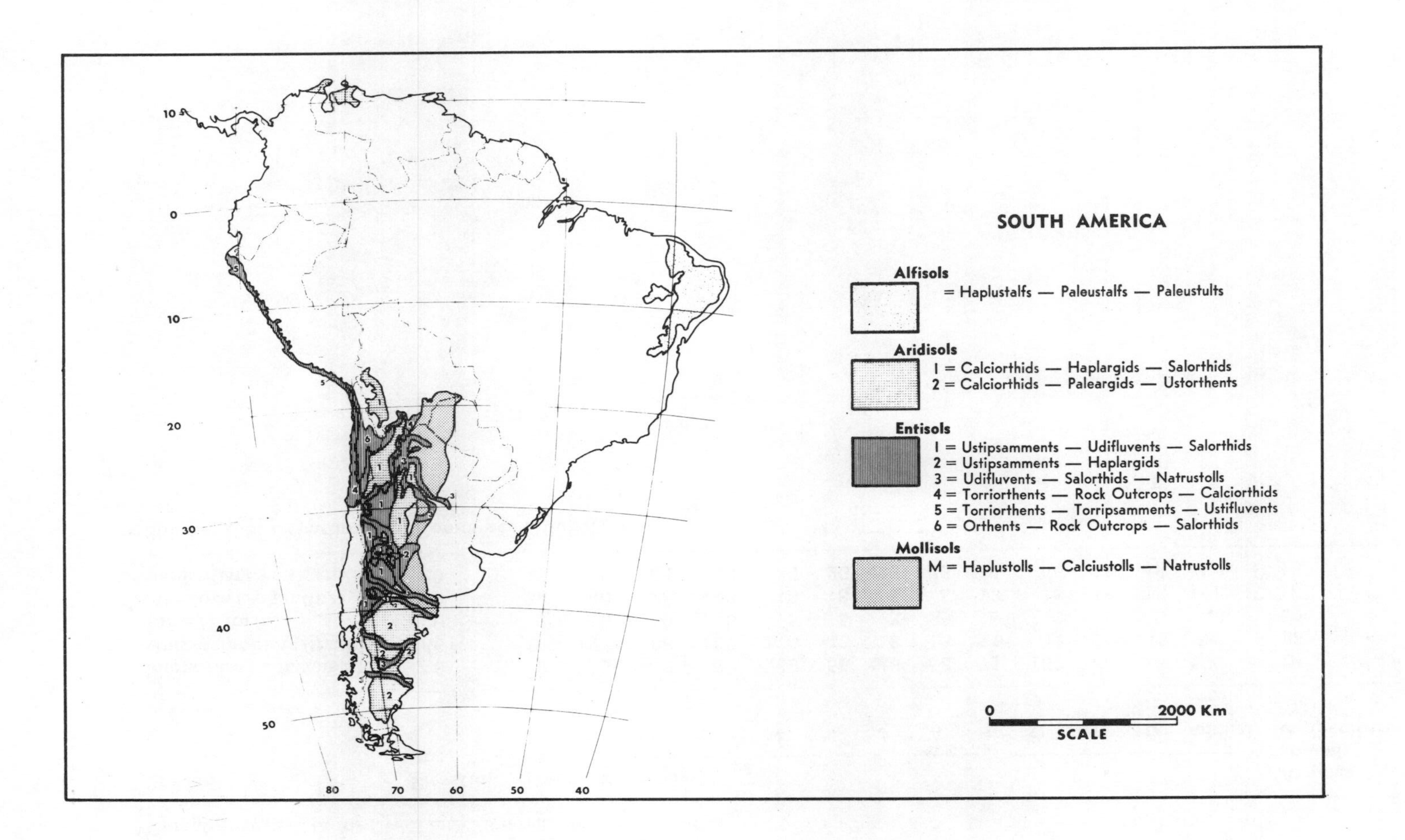

Fig.8.2. Arid-region soils of South America.

TABLE 8.1

Climatological data for five South American weather stations

Station	Elevation (m)	Precipitation (mm)													Average annual temperature (°C)
		J	F	M	A	M	J	J	A	S	O	N	D	annual	
Maracaibo (Venezuela)	6	2	0	8	20	69	56	46	56	71	150	84	15	577	29
Quixeramobim (Brazil)	166	18	127	168	127	178	43	18	15	10	15	18	15	752	28
Lima (Peru)	120	2	0	0	0	5	5	8	8	8	2	2	0	40	20
Mendoza (Argentina)	800	23	30	28	13	10	8	5	8	13	18	18	18	192	16
Bahia Blanca (Argentina)	29	43	56	64	58	30	23	25	25	41	56	53	48	522	16

Source: U.S. Department of Commerce (1969).

Quixeramobim is wet by arid-region standards but the effectiveness of the rain is low because it comes mostly when temperatures are high. Northeastern Brazil resembles the Sahel in sub-Saharan Africa in both the precipitation patterns and the kinds of soils it has. Precipitation on the Altiplano of Bolivia generally averages less than 200 mm and the region abounds in saline depressions (salares).

Annual temperatures decrease from north to south but the temperature range increases in the same direction. At Maracaibo, the maximum temperature was 39°C and the lowest 19°C, whereas at Bahia Blanca the highest was 43°C and the lowest a minus 8°C. Mendoza has about the same maximum and minimum temperatures as Bahia Blanca. The growing season for crops is long everywhere except in the far south and in the high mountains.

Geology and geomorphology

The oldest rocks in South America form the basement complex for the Patagonian Desert and the arid section of northeastern Brazil. They consist of metamorphosed Precambrian rocks of what are called the Extra-Andean Patagonian shield and the Brazilian shield. Landscapes in the Brazilian shield are characterized by level erosion surfaces of different ages, giving a stepped appearance to the land. While the overall aspect is that of a plain, the landscape is actually rather dissected due to the number of small and large valleys, each containing one or more pediplains. Because the surfaces are old, themselves, or are the eroded sediments from old surfaces, the soils are highly weathered and relatively infertile.

While the basement rocks of the Extra-Andean Patagonian shield are Precambrian, as in the Brazilian shield, large depressions filled with marine and continental beds of sedimentary rocks are found between the massifs where Precambrian rocks are exposed. Volcanic rock flows and tuffs form important admixtures with the marine and continental sediments. The topography is level to undulating in the depressions, with several major valleys cutting across them from west to east. The mountains are relatively low and rounded.

Rocks in the Andean System — stretching the entire length of the west side of the continent — vary widely in age and composition. Tectonic activity has continued for a long time and has resulted in the development of many depressions (now filled with sediments) interspersed between the more or less parallel eastern and western mountain ranges making up the Cordillera de los Andes. Additionally, many volcanoes remain active and are responsible for periodic lava flows and the deposition of volcanic ash on the landscapes. Several glaciations during the Pleistocene have influenced the type of sediments occupying the depressions. The Andean relief, as a whole, is youthful and unstable.

In the more arid parts of the main Andean system (the northwest coast on the Caribbean, the coastal plain of Peru and Chile, and the Altiplano centered in Bolivia) the topography is planar. Several marine terraces are present along the coast from southern Ecuador to northern Chile. Pediments are widespread in the Altiplano, indicating that active erosion has occurred in the past during more humid periods of the Pleistocene. The Altiplano itself is a very large closed basin containing a number of subbasins where numerous salt flats, the largest of which is the Salar de Uyuni, and two large lakes (Titicaca and Poopó) occur. In the pampean region of northwestern Argentina, the planar topography is broken by mountains composed of Precambrian rocks and Quaternary sediments.

East of the Andes, the level land surface slopes gradually toward the Atlantic Ocean. The only major break in the plain occurs where the Sierra de Córdoba forms a humid island west of the city of Córdoba. Broad, flat depressions, frequently having saline or sodic soils, dot the planar landscape. Sediments of Tertiary and Quaternary age, with admixtures of volcanic ash and other aeolian materials, make up the parent material for the soils.

Vegetation

Open and closed savannahs represent the principal kind of vegetation systems found on all the Alfisols and on the Mollisols of the Gran Chaco of Argentina, Paraguay, and Bolivia. Vegetation on the Aridisols and Entisols consists of shrubs combined with a few ephemeral plants in the driest places and with perennial grasses in the more favored areas. Mollisols of the great pampas of Argentina are covered with perennial grasses.

In the small arid region of northern Colombia and northwestern Venezuela, cacti (*Cereus*, *Cephalocereus*, *Opuntia*) and thorny woodland (*Acacia*, *Mimosa*, *Prosopis*) species occupy the drier areas while trees such as *Curatella* and *Bursonima*, with *Panicum*, *Paspulum*, and *Azonopus* grasses, are common in the wetter sections.

A vegetation formation known as caatinga dominates the uplands of northeastern Brazil. It consists of cacti and thorny woodlands similar to that found in the arid regions of Colombia and Venezuela. *Cereus*, *Cephalocereus*, and *Pilocereus* are the principal cacti; *Caesalpina*, *Cavanillesia*, *Mimosa*, and *Acacia* are common thorny trees. The vegetation is xerophytic despite the rather high rainfall, which is highly variable from year to year and is concentrated in the five months from January to May. There are two types of savannahs occupying level terrain in that same region: campo cerrado (tall grasses with scrubby trees) and campo limpo (treeless grassland). The savannah grasses include *Aristida*, *Eragrostis*, *Paspalum*, and *Panicum*, and the trees are mainly *Caryocar*, *Curatella*, and *Aspidosperma* species.

The extensive south and west coast arid regions have five distinctive vegetative formations. There are the grasslands of the Argentinian pampa, the savannahs and woodlands of the Chaco, the steppes of Patagonia and western Argentina, the montane desert of the Altiplano, and the coastal desert of Chile and Peru.

In the arid western portion of the pampa grasslands, *Stipa* is by all odds the dominant grass, with some *Festuca* and *Sporobolus* also present in significant amounts.

Algarrobo (*Prosopis* spp.), quebracho blanco (*Aspidosperma*), and quebracho colorado (*Schinopsis*), along with some *Zizyphus* and *Caesalpina*, are the principal trees in the Chaco savannahs and woodlands. The understory of the savannahs consists mainly of cacti (*Cereus*, *Opuntia*) and xerophytic shrubs (*Acacia*, *Mimosa*). Since the Chaco has a low-lying level terrain traversed by shallow rivers, flood waters originating in the north and west commonly inundate vast sections of land. Swamps and salt flats are widespread.

The bushy steppe of western Argentina and Patagonia contains a mixture of tussock grasses (*Stipa*, *Festuca*) and low thorny shrubs (*Larrea*, *Cassia*, *Verbena*) and, occasionally, *Prosopis*. The vegetation becomes more grassy and less shrubby in the south of Patagonia where precipitation effectiveness is high due to the low temperatures. In the latter areas, the dominant grasses are *Festuca*, *Poa*, and *Agropyron*. There are significant areas of saline and sandy soils in the drier part of the zone where halophytes such as *Atriplex* and psammophytes such as *Panicum* and *Plazia* are the main species.

Low shrubs represent the major vegetation type in the montane desert of the Altiplano at an elevation of 3,000 or more meters. Most of the montane desert lies in Bolivia, with lesser amounts in Argentina, Chile and Peru. Between the widely scattered shrubs and the tufts of grass will be bare soil or a few ephemerals, usually. The principal shrub species are *Lepidophyllum*, *Baccharis*, *Verbena*, and *Pseudobaccharis*. Among the sparse grasses are *Stipa*, *Eragrostis*, *Panicum*, and *Bouteloua*. There is some *Opuntia* cactus. Halophytes (*Salicornia*, *Distichlis*) occupy the perimeter of the numerous saline flats (salares), the centers of which are usually barren when dry. In places where there is a continuous or nearly continuous small supply of running water from springs or ice fields, the unique vegetation type known as *bofedales* is found. It consists of bogs of *Ephedra* and *Atriplex*, mainly, and is the primary grazing area for llamas and alpacas of the Altiplano.

The fifth distinctive vegetative formation is the coastal desert where little or nothing grows. The "vegetation" may be only a few soil microorganisms or primitive *Tillandsia*. Herbaceous communities or lichens and woody plants are absent unless there is a supplemental source of water, as along streams debouching from the Andes. Stunted trees of *Carica*, *Caesalpina*, *Acacia*, and *Eugenia* occupy some of the low hills (lomas) between the Andes and the Pacific Ocean. The trees are dependent for survival upon capture of moisture from the fogs which come in from the cold Pacific.

SOILS

Arid-region soils cover about 16% of South America (Table 8.2). Entisols of alluvial fans, piedmonts, and flood plains make up most of the Entisol areas in Argentina and Bolivia, whereas shallow soils on marine deposits are dominant in Chile and Peru. Arid-region Alfisols are confined to northeastern Brazil where annual precipitation is relatively high, summer is the wet season, and droughts are frequent. Aridisols are dominant in the hot Caribbean coast of Colombia and Venezuela and in western Argentina, as well as in the cool Altiplano and Patagonia.

The southern part of the Mollisol area is the western extension of the Argentine pampa grasslands while the northern part is in the Gran Chaco woodlands and savannahs. Saline soils are extensive in the southern part of the continent and sodium-affected soils are common in semiarid Argentina, Bolivia, and Paraguay.

Alfisols

As in Africa, Alfisols of South America are the soils of the xerophytic deciduous forests and savannahs where the plant growing season is long, temperatures are rather high, precipitation is heavy enough during some part of the year to cause considerable leaching of the soils, and the dry season persists for several months. On the wetter side of the Alfisol zone, the soils grade into Ultisols.

Alfisols of northeastern Brazil are at least moderately well drained throughout the profile even though they may be fine textured in the surface. They possess a well-developed argillic horizon which is not markedly different in texture from the surface soil, has a base saturation of 50% or more, and contains significant amounts of sesquioxides.

TABLE 8.2

Area of dominant soil orders in arid regions of South America

Soil order	Area (km^2)	Percent of	
		arid region	continent
Alfisol	380,000	13.4	2.2
Aridisol	790,000	27.9	4.5
Entisol	1,175,000	41.4	6.7
Mollisol	490,000	17.3	2.8
Total	2,835,000	100.0	16.2

A1. Haplustalfs — Paleustalfs — Paleustults

Alfisols of northern Brazil approximate Ultisols in their properties. Soil fertility is moderate, land surfaces are old, and the soils are highly eroded. The normal dry season is long and droughts are frequent even though the mean annual precipitation is more than 500 mm. Due to the long dry season, the vegetation consists largely of the caatinga type, with cacti and thorny woodland dominant. The principal clay minerals probably are kaolinite and illite, with some montmorillonite in the lower-lying areas. Soil parent material consists of metamorphic (gneiss, schist, quartzite, slate) and sedimentary (limestone, sandstone) rocks. The topography is rolling to hilly.

Haplustalfs are moderately deep soils on moderate slopes, usually are medium to fine textured in the surface but sometimes sandy, have a dark reddish-brown or reddish-brown surface-soil color, and are low in organic matter. The argillic horizon abruptly underlies the A horizon. Permeability to water is generally good. The surface soil is hard to very hard when dry and the red subsoil hardens almost to the point of being classed as plinthite. Base saturation ranges from 50—80%, in the main, and sometimes there is an accumulation of carbonates at the base of the argillic horizon or in the upper part of the C horizon.

Paleustalfs are deep reddish soils on level terrain containing thin and discontinuous plinthite horizons in the subsoil. Organic-matter content is low, base saturation is on the order 40—70%, soils are acid throughout the solum, and there is no accumulation of calcium carbonate within 2 m of the surface. Surface soils are medium textured, usually, and are very hard when dry. There is an abrupt textural change between the A horizon and the argillic B horizon. Permeability to water is fair to good.

Paleustults are, like the Paleustalfs, deep reddish soils and are the only Ultisols included among the arid-region soils. The plinthite horizons are thicker than in the Paleustalfs but still are discontinuous near the surface, base saturation is lower (less than 35%), and cation-exchange capacity is low due to the dominance of kaolinitic and hydrous-oxide clay minerals. The soil usually has a sandy loam texture but the surface may be a sand or loamy sand. Paleustults occupy old dissected surfaces. Soil reaction is moderately to strongly acid. Paleustults are found most often in the humid fringe of the Alfisol zone and are intermingled with Paleustalfs.

Other great groups in this association include Ustorthents, Udifluvents and Chromusterts.

Aridisols

Geographically and topographically, Aridisols of South America cover a wide range, from the hot northern tip of the continent almost to the cold southern tip and from sea level to 5,000 m. Weak development is the distinguishing characteristic of the majority of Aridisols. Calciorthids and

Salorthids are the dominant great groups, and neither shows much evidence of development. Within the Aridisol zones, the Entisols are the next most prevalent soil order. The latter are either shallow soils (Orthents) or alluvial soils (Fluvents), mainly. Salinas and salares are absent or rare in the north of the continent but are numerous and large in western Argentina, southwestern Bolivia, northern Chile, and along the Atlantic Ocean in Patagonia.

AR1. Calciorthids — Haplargids — Salorthids

Aridisols in this association have in common a desert shrubland type of vegetation, with cacti and thorny bushes or small trees and a sparse grass cover. Elevations range from the 4,000 m Altiplano of Bolivia to the 30 m lowlands of the Guajira Peninsula in Colombia and Venezuela. The topography is smooth to undulating except in northwestern Argentina, which has a basin-and-range landscape. Precipitation varies between about 200 and 500 mm and is seldom high in any one month. Large salt lakes, called salares and salinas, cover much land area in the Altiplano and in Argentina. Barren shifting sand dunes are common in the Guajira and are found less commonly in the other areas. The dominant clay minerals in the Guajira appear to be kaolinite and hydrous oxides whereas montmorillonite and allophane probably are the principal ones in the Aridisols of Bolivia and Argentina.

Calciorthids are the Aridisols without an argillic or natric horizon but having a calcic horizon within 1 m of the surface. The soils are coarse textured, generally. They frequently are calcareous throughout the profile although the surface soil, if coarse textured, may be noncalcareous in the upper several centimeters. Base saturation is high. Subsoils sometimes are saline, particularly in soils near salt flats. Soil reaction is neutral to moderately alkaline. Organic-matter content is less than 1% and the soils are moderately deep to deep. Soil color is grayish brown to reddish brown. Calciorthids usually are found on the more sloping land. Due to their sandy surface texture, they are susceptible to wind erosion.

Haplargids in this delineation have a well-defined argillic horizon which usually is calcareous in its lower part. Surface-soil texture mainly is sandy loam or loam and the B horizon is a sandy clay loam or clay loam. The soils are shallow or moderately deep and low in organic matter, with a slightly acid soil reaction in the surface and a slightly to moderately alkaline argillic horizon. Base saturation is high. They occupy level to gently sloping upland areas. Some of them are covered with a desert pavement.

Salorthids are the salty soils of the salares and salinas, with a water table within 1 m of the surface for one or more months. The salt lakes are impressively large: the Salinas Grandes in Argentina are over 150 km long and 50 km wide and the Salar de Uyuni in Bolivia covers an even larger area. Vegetation is sparse or nonexistent in the salt flats and wind erosion whips up clouds of salt crystals during the dry season. Soils are fine textured, show little development, may have gypsum crystals in the subsoil, frequently are calcareous, are slowly permeable to water, and have a moderately alkaline reaction. Polygonal cracking shows up on the surface of soils in the slight depressions within the flat terrain.

Torripsamments, Torriorthents, Ustifluvents and Natrargids are other great groups in this association.

AR2. Calciorthids — Paleargids — Ustorthents

Patagonia, in southern Argentina, is the locale for this group of Aridisols and Entisols. The landscape consists of low mountains, dissected tablelands, terraces, and plains on aeolian, fluvial, and lacustrine deposits with admixtures of volcanic ash nearly everywhere. Tableland soils are shallow and stony, with a desert pavement cover, whereas the terraces and plains have moderately deep and deep soils. Annual precipitation ranges from 150 to 300 mm and is fairly evenly distributed through the year, and temperatures are low. Winds are strong, especially in spring, and wind erosion is a major problem in some areas. The vegetation is of the bushy-steppe type with scattered tuft grasses (*Stipa* and *Festuca*) and low thorny shrubs (*Mulinum, Verbena, Ephedra*). In the colder and wetter south the steppe contains more grass and less shrubs.

Calciorthids of Patagonia have a calcic horizon at a rather shallow depth (usually less than 30 cm) and frequently are noncalcareous in the upper few centimeters. Surface soils generally are medium textured fine sandy loams, loams, or silt loams, but there are some coarse-textured soils, as well. Subsoils may be somewhat finer textured and have a more developed structure than the surface soils. Occasionally, the calcic horizon is indurated and the soils would be classed as Paleorthids. Soil reaction is moderately alkaline although the surface horizon may be acid. Organic-matter content is low. The dominant colors are brown and grayish brown. Soils are moderately deep to deep and the topography is level to undulating on the terraces and plains where these soils occur.

Paleargids in this association have an argillic horizon which abruptly underlies the A horizon. Surface textures are usually loams, whereas the argillic B horizon is a clay loam. The latter horizon is strongly developed and plastic. While the surface color commonly is a brown or dark reddish brown, the argillic horizon is a reddish brown. Permeability to water is moderate to slow. Soil reaction is slightly acid throughout the solum, organic-matter content is low, and the soils are moderately deep to deep. They occupy level surfaces of old tablelands and terraces.

Ustorthents are the shallow stony soils of the low mountains and tableland slopes. Textures vary from coarse to medium, grayish brown is the usual color, organic-matter content is low, the topography is steep, rock underlies the soils at less than 20 cm, and rock outcrops are common.

Other great groups of significance in this association are Ustipsamments, Ustifluvents, Udifluvents, Natrustolls, and Salorthids.

Entisols

Orthents, Fluvents, and Psamments are all important members of the Entisol order in arid South America. Orthents occupy a good part of the extremely arid coastal plain of Chile and Peru, the higher parts of the gently undulating plain of western Argentina, the Andes mountains, and an area in northeastern Brazil. Fluvents are found everywhere, sometimes in small drainageways and other times in broad stream valleys. Psamments are of greatest extent in central Argentina. In all Entisols, soil development is minimal. The dominant land use on the Orthents is grazing, Fluvents are the principal irrigated soils, and Psamments are used for grazing in the more arid regions and for dryland farming in the semiarid regions. Salt-affected soils occur in all the Entisol areas. Silicate minerals are dominant and volcanic ash is a component of most soils.

E1. Ustipsamments — Udifluvents — Salorthids

Soils of this association are found on the western side of the great Argentina plain where the parent material consists of sedimentary deposits of fluvial and aeolian origin. Average annual precipitation is less than 300 mm and crop production is possible only when irrigation water is available, as it is in the valleys heading in the Andes. Summer temperatures are high. The vegetation is dominated by *Prosopis* and *Larrea*. The principal grasses are *Stipa* and *Festuca*, but they afford only a sparse cover. Elevation differences between the valley bottoms and the low hills usually amounts to only about 10—20 m.

Ustipsamments in Argentina are moderately deep loamy sands on fairly level upland surfaces, with small dunes present in places where the soil surface has been disturbed. Textures are uniform down to the underlying sedimentary deposits. The structure is massive, for the most part. Soil reaction is neutral to slightly alkaline, organic matter is low, salinity is very low, and there is no horizon of calcium carbonate accumulation although the soils are slightly calcareous. The dominant color is reddish brown.

Udifluvents are recent alluvial soils in broad valleys crossing the plain in a mostly west—east direction. The soils are deep and somewhat stratified, but the surface-soil texture is mainly loam to clay loam. Organic-matter content is moderate; the soils are slightly to moderately alkaline and are calcareous in the subsoils; most of them are slightly to moderately saline; and soil color is yellowish brown to light brown. Permeability to water usually is moderate to good, but some soils are slowly permeable.

Salorthids are deep soils found in level closed basins, often very large in area, and in the irrigated valleys where water tables are close to the surface. In the closed basin where salt flats (salinas) occur, the soils are fine textured on the surface but may have lenses of sandy material in the subsoil. Permeability is slow to very slow. Although the basins are flat, there is a micro-relief of small depressions. Polygonal cracks appear in the soils of the depressions, whereas salt crusts are common on the slightly higher places between the depressions. The water table fluctuates between 1 and 2 m below the surface and the profiles are saline except in the top few centimeters in the depressions. Many of the salina soils apparently contain considerable amounts of exchangeable sodium. Surfaces are usually barren of vegetation or have a sparse cover of halophytic forbs and shrubs. Salorthids in the irrigated areas are the result of high water tables caused by canal seepage, heavy irrigation, and inadequate drainage, leading to abandonment of the soils insofar as crop production is concerned. Soil salinity is very high in the surface and some of the soils show signs of the black color associated with calcium and magnesium chlorides.

Other important soils are Calciorthids and Ustorthents on the upland sites.

E2. Ustipsamments — Haplargids

While sandy soils are widespread in western Argentina, one particular area in La Pampa, San Luis, and Cordoba provinces consists almost entirely of sandy soils, and wind erosion and the stabilizing of dunes is a major problem. In this semiarid region where the average annual precipitation is between 400 and 600 mm, destruction of the native grass and shrub cover when dryland farming is practical has led to the formation of barren, moving sand dunes in so much of the area that they now represent a serious problem for highways, railroads, and villagès, as well as on the land itself. Effective techniques for reestablishing a vegetative cover on the dunes have been developed but the economic returns from dune control have not been high enough to induce land owners to invest in the practice. *Prosopis* is the dominant native shrub in the sandy lands and it, with *Populus* and *Eucalyptus*, is among the shrubs used for dune stabilization. Within the expansive plain where the sandy soils occur, Ustipsamments are found on the dunes and Haplargids between the dunes.

Ustipsamments probably occupy much less than half of the region, but they appear to dominate the area because barren dunes are so prominent and the other great group, the Haplargids, has a sandy surface soil which usually presents a hummocky appearance under rangeland conditions. The hummocks and dunes marking the location of Ustipsamments vary in height from as little as 50 cm to as much as 20 m, with the average probably between 1 and 2 m. Wind sorting has removed most of the silt, clay, and organic matter in the surface of the original Haplargids, leaving the dunes with 80% or more fine and very fine sand and 95% or more of total sand. Soils are deep, in the main, noncalcareous, neutral to slightly alkaline, very low in organic matter, and have little or no structural development. They are effective trappers of moisture because they are highly permeable and rapidly form a dry surface which greatly reduces evaporative losses from the subsoil.

Haplargids are characterized by a well-developed argillic horizon beginning at 20—30 cm below the soil surface. Surface-soil texture is a fine sandy loam or a loamy fine sand; subsoils are sandy clay loams, typically. The argillic horizon has a blocky structure and is at least moderately permeable. Soil reaction is slightly acid to slightly alkaline on the surface and moderately alkaline in the subsoil. The soils commonly are nonsaline and are noncalcareous above the argillic horizon and calcareous in and below that horizon. The dominant color is gray-brown in the surface and yellowish brown in the lower part of the profile.

Among the great groups present to a minor extent are Udifluvents, Salorthids, and Natralbolls.

E3. Udifluvents — Salorthids — Natrustolls

Intermingled saline and nonsaline alluvial soils make up most of this association. The valleys are broad, flood plains extending from one side of the valley to the other, rivers are shallow, floods are common, and water tables may be close to the surface. Only the soils in the flood-plain depressions (Natrustolls) show significant signs of development. Average annual precipitation is between 100 and 700 mm and the dominant native vegetation consists of *Prosopis* species able to survive periodic inundations, salinity, and high water tables.

Udifluvents are the flood-plain soils of the broad valleys of rivers such as the Bermejo, Salado, and Negro. Stratification is weak, most of the layers being some variation of medium textured. Where fine-textured strata occur in the subsoil, the soils usually are Natrustolls rather than Ustifluvents. Organic-matter content is moderate, salinity is slight to moderate, soil reaction is slightly acid to slightly alkaline in the surface, the lower subsoil may be calcareous, and permeability is fair to good. In the drier west, the Ustifluvents are irrigated and are productive when properly managed.

Salorthids are problem soils because of their high water tables and high salinity. Their properties are similar to those of the Udifluvents, with the exception of their higher salinity. The principal source of the salt appears to be the underlying sedimentary deposits rather than the salt carried by the rivers. The water table frequently lies within 1 m of the surface.

Natrustolls are sodium-affected dark-colored Mollisols and are found in slight depressions in the valley plains. Organic-matter content of the surface soil exceeds 2%, the surface soil is medium textured, the natric subsoil horizon is fine textured and slowly permeable, soil reaction is moderately alkaline in the surface and more strongly alkaline in the subsoils, the soils are wet for several months of the year and commonly are at least moderately saline in the natric horizon. The natric horizon has a columnar structure and may be calcareous in its lower part. Some mottling can be detected in many of the subsoils.

There also are Argiaquolls in this association.

E4. Torriorthents — Rock Outcrops — Calciorthids

Shallow and very shallow soils constitute the major part of this association of mountains, hills, and rather narrow valleys along the flanks of the Cordillera de los Andes and the Cordillera de la Costa. Vegetation is scarce everywhere and is nearly nonexistent on the Chilean coast. In Chile, many of the upland soils are saline due to cyclic salts and the salinity of the parent material. *Prosopis* (tamarugo in Chile; algarrobo in Argentina) is the principal shrub and tree in this association, as it is in so much of western and northern Argentina. *Larrea* is the shrub of secondary importance in Argentina. Annual precipitation is less than 100 mm in the west and south but increases to about 300 mm in the northeast. Temperatures are high on the east side of the Andes but are lower on the west side.

Torriorthents are skeletal soils overlying hard rock and soft sedimentary deposits. The topography is moderately or steeply sloping, surface rocks are common, there is little evidence of any organic-matter accumulation in the surface soil, most of the soils are calcareous and those along the coast of Chile are saline and frequently gypsiferous, and the dominant color is yellowish brown.

Rock Outcrops is a land form consisting of surface stones and exposed underlying rocks. There are a few pockets of very shallow soils between the rocks. This land form is found on the steeper slopes.

Calciorthids are the only soils of any significant depth in this association. They occupy the more gently sloping fans at the foot of the mountains. Their distinctive horizon is the calcic horizon which begins at 20—30 cm. The soils are coarse textured, frequently gravelly, moderately deep to deep over bedrock, low in organic matter, calcareous to the surface, and show little evidence of development. There sometimes is a petrocalcic (indurated) horizon of calcium carbonate and silica.

Other great groups in this association include Ustifluvents and Salorthids.

E5. Torriorthents — Torripsamments — Ustifluvents

The narrow coastal plain of Peru, where most of the nation's population lives, is the location of this group of soils. The landscape consists of: (1) a level plain where undeveloped soils (Torriorthents) and deep sands (Torripsamments) occur; (2) low hills rising from the plain, again with undeveloped soils (Torriorthents); (3) unstable sand dunes of moderate size and extent (Torripsamments), principally along the coast; and (4) some 52 valleys running from the Andes to the Pacific Ocean, where productive alluvial soils (Ustifluvents) are used for irrigation (Comité Peruano de Zonas Aridas, 1963).

Average annual precipitation ranges from less than 10 mm in the south to nearly 300 mm in the north. In the southern and central part of the coastal plain, there is no significant vegetative cover except at the 200—800 m level on the ocean-facing side of hills, where moisture condensing from winter fogs supports the sparse lomas vegetation of lichens, grasses, and scattered shrubs and trees (Cornejo, 1970). In the more humid north, *Prosopis* and *Copparis* are dominant but there is some cactus (*Cereus*).

Torriorthents are the most extensive soils on the Peruvian coastal plain. They are shallow to deep soils, dominantly medium textured, show little or no development, are very low in organic-matter content, have a neutral to slightly alkaline reaction, are generally calcareous and at least slightly saline (and sometimes highly saline), have a yellowish-brown to light-brown color, and frequently are covered with a desert pavement. Torriorthents are found on both the level coastal plain and on the coastal mountains lying between the Andes and the Pacific Ocean. The shallow soils occur most commonly on the tops of hills and on the marine terraces composed of fine-textured sediments. The deep coarse to medium textured Torriorthents have considerable potential for crop production if irrigation water is available.

Torripsamments are predominantly of two types here: (1) moderately deep sandy soils on a very gently undulating plain in the northern coastal area and (2) deep sands of unstable barren dunes next to the ocean all along the coast. Additionally, there are sand dunes on the interior plain in the south and the north. The sandy soils are light brown to reddish brown in color, slightly acid to neutral in reaction, may be saline in the subsoil, are generally noncalcareous in the upper part of the profile, and are low to very low in organic matter. Since annual precipitation increases from south to north, the amount of vegetative cover increases in the same direction but it is always sparse, at best.

Ustifluvents are the alluvial soils in the many valleys extending across the coastal plain. Virtually the entire population of western Peru is concentrated in the valleys where irrigated agriculture is practiced. The valleys are narrow in the Andes but widen out as they cross the coastal plain, attaining a width of 5—20 km when they reach the ocean. Their length after they leave the Andes is about 50—75 km. Irrigation water comes directly from the rivers or from wells in the valley fill. The soils are deep, highly stratified and variable in texture, with enough fine-textured layers present to pose a drainage problem when excessive amounts of irrigation water are applied to the fields. Salinity was a minor problem originally but has assumed major importance in recent years due to the development of perched water tables in the irrigated sections. Soil reaction varies from moderately alkaline to strongly acid, depending largely upon the pH of the river and well waters. Boron levels are high in many of the irrigated soils.

The other important great group is Salorthids, most of them man-made.

E6. Orthents — Rock Outcrops — Salorthids

Salinas (salares), barren rock, and shallow soils over rocks and sedimentary deposits constitute practically all of the surface materials in this delineation located in northern Chile and northwestern Argentina. Salinas and salares are salt flats; the names are interchangeable. The region is an inhospitable one where there is absolutely no vegetation in some parts and very little anywhere. What vegetation there is consists of *Prosopis* and *Acacia* trees and some shrubs. Elevations range from about 1,000 m to over 5,000 m. Aridity is extreme but, paradoxically, high water tables lead to formation of salt flats in the depressions. Most of the water, which is saline, comes underground from the nearby and better-watered high mountains. The landscape consists of mountains, low hills, and broad valleys.

Orthents and *Rock Outcrops* are intermingled, the shallow Orthents grading into Rock Outcrops over much of the area. Soil formation is so slow where rainfall is practically nonexistent that the soil may consist of only a few millimeters of weathered material. Development is greater at the higher elevations than it is in the lowland desert of Chile. Orthents occupy much of the large and broad valley between the Cordillera de los Andes and the Cordillera de la Costa. The valley landscape is smooth or undulating, in the main, but does contain small rounded hills. Most of the valley plain consists of coarse-textured sedimentary deposits and much of the material is saline. The major nitrate deposits are in this association.

Salorthids have a water table within one or two meters during much of the year, are highly saline in the surface horizon and at least saline below that horizon, show little evidence of development, are fine textured, have water standing on the surface for varying periods of time after runoff from the higher elevations, and occupy broad flat depressions. Organic-matter content is low and the soils are moderately alkaline.

Among the other great groups are Torrifluvents and Calciorthids.

Mollisols

M. Haplustolls — Calciustolls — Natrustolls

Mollisols of various kinds are present along the east side of the arid zone in southern South America and — to a small degree — in the wetter parts of the Altiplano. Mollisols of the semiarid Argentine pampas are transformed gradually into the Mollisols of the subhumid zone as one goes from west to east across the great plain which extends from the Andes to the Atlantic Ocean. In the Gran Chaco, Mollisols grade into Alfisols on the east. The ubiquitous *Prosopis* (algarrobo) is the dominant woodland species throughout this association. In the Chaco, *Aspidosperma* and *Schinopsis* are important quebracho trees; the former is white quebracho and the latter is red quebracho. *Stipa* and *Festuca* are the main grasses. Average annual precipitation is about 400—700 mm. Annual temperatures are higher in the north than in the south. Volcanic ash is present in all the soils.

Haplustolls have a thin dark-brown surface soil, contain considerable organic matter, are medium textured in the surface and somewhat finer textured in subsoil where a cambic horizon commonly occurs, have a very high base saturation beneath the surface horizon and are calcareous below about 50 cm, and have a slightly to moderately alkaline reaction. Many Haplustolls are close to being Argiustolls. Permeability is moderate to good, generally. The soils are productive when irrigated and leached of salt. Without irrigation, they are used mainly for grazing in the north and for dryland farming and grazing in the south.

Calciustolls are the Mollisols having a calcic or petrocalcic horizon. In Argentina, petrocalcic horizons cemented with calcium carbonate and silica are known as tosca. Petrocalcic horizons are common in the soils of the southern part of the association; they are less common in the north. Typically, Calciustolls have a fairly thick A horizon overlying the calcic or petrocalcic horizon and usually are calcareous throughout the profile. There frequently is a cambic horizon above the calcic or petrocalcic horizon. Surface texture is sandy loam; subsoils are finer textured. The soils are moderately deep to deep.

Natrustolls occur in the slight depressions which pockmark the Argentine pampa and the Chaco. They are dark-colored sodium-affected soils having the natric horizon within 50 cm of the surface, normally. The soils are medium to fine textured in the surface and fine textured in the subsoil. Soil reaction is moderately alkaline in the surface and strongly alkaline in the subsoils; salinity is usually high. Subsoils may be calcareous in the lower part of the natric horizon. Permeability is low.

Argiustolls and Ustipsamments are common and there are some Ustifluvents and Salorthids.

Salinity

One of the striking features about the soils of the arid part of southern and western South America is the occurrence of saline soils everywhere, from Patagonia to the pampas, the Chaco, the Altiplano, and the coastal plain of Chile and Peru. There are more salinas and salares in western Argentina and Bolivia, per unit area, than anywhere outside the Dasht-i-Kavir salt desert in Iran. Additionally, the Atacama Desert soils are nearly invariably saline, as is a large percentage of the western and southern Chaco soils. Salinas are common in Patagonia, and sodium-affected soils are widespread in the depressions of the pampas. Faulty irrigation practices have led to salinization of much irrigated land, especially in Peru and Argentina, and the salinized area continues to expand.

Most of the salt undoubtedly comes from the marine intrusions into what is now Argentina, Paraguay, and Bolivia and by uplifts of marine sediments associated with the formation of the Cordillera de los Andes. Weathering of the young Andean rocks and subsequent erosion have contributed to the salt problem in valleys heading in the Andes and in Andean closed basins. Present additions of cyclic salt from both the Atlantic and Pacific oceans probably are significant at the lower elevations.

All in all, salinity can safely be said to be the number one edaphic problem in the arid regions of South America.

REFERENCES

Comité Peruano de Zonas Aridas, 1963. *Informe Nacional sobre las Zonas Aridas.* República del Perú, Lima.

Cornejo, T.A., 1970. Resources of arid South America. In: H.E. Dregne (Editor), *Arid Lands in Transition. Am. Assoc. Adv. Sci., Publ.* No. 90: 345—380.

Etchevehere, P.H., 1971. *Mapa de Suelos.* República Argentina, Centro de Investigaciones de Recursos Naturales, INTA, Castelar.

FAO/UNESCO, 1971. *Soil Map of the World, 1:5,000,000, Volume IV, South America.* Unesco, Paris, 193 pp., 2 maps.

Meigs, P., 1953. World distribution of arid and semi-arid homoclimates. In: UNESCO, *Reviews of Research on Arid Zone Hydrology. Arid Zone Res.*, I: 203—210.

Papadakis, J., 1963. Soils of Argentina. *Soil Sci.*, 95: 356—366.

U.S. Department of Commerce, 1969. *Climates of the World.* Environmental Data Service, Environmental Science Services Administration, U.S. Department of Commerce, Washington, D.C., 28 pp.

Chapter 9

SPAIN

INTRODUCTION

Spain is one of two European countries having arid areas; the other is the Soviet Union. The description of the arid European part of the U.S.S.R. is included in the chapter on Asia.

Somewhat more than 50% of Spain is arid, all of it falling in Meigs' semiarid category (Fig.9b) although there may be places in the southeast where the climate is drier than semiarid. Aridisols and Entisols are the dominant soil orders (Fig.9a). Most of the soil information came from the book by Villar (1937).

ENVIRONMENT

Climate

All of Spain has a Mediterranean climate with moist winters and hot, dry summers. Summer dryness is less pronounced in the north than in the south part of the country. Precipitation effectiveness is high because rains come in the cool period of the year but irrigation is essential for maximum yields of summer crops.

Climatological data for three weather stations in Spain are given in Table 9.1. Burgos is in the northern tablelands, Madrid in the center, and Almería on the southern Mediterranean coast. Almería is considerably more arid than the remainder of the country, it had the highest maximum temperature of the three stations (42°C), and frosts do not occur there. Burgos and Madrid had maximum temperatures of 37°C and 39°C and minimums of −18°C and −10°C, respectively.

Geology and geomorphology

Calcareous and gypsiferous formations are widespread in the arid regions. Limestone, sandstone, and shale are the principal rocks in the northern and southern tablelands and in the old marine basins of the Ebro and Guadalquivir valleys. Igneous rocks are present in the mountain ranges which rise to over 3,500 m in places. The tablelands have an undulating to rolling topography, as do the uplands along the Ebro River. Flood plains of the numerous rivers

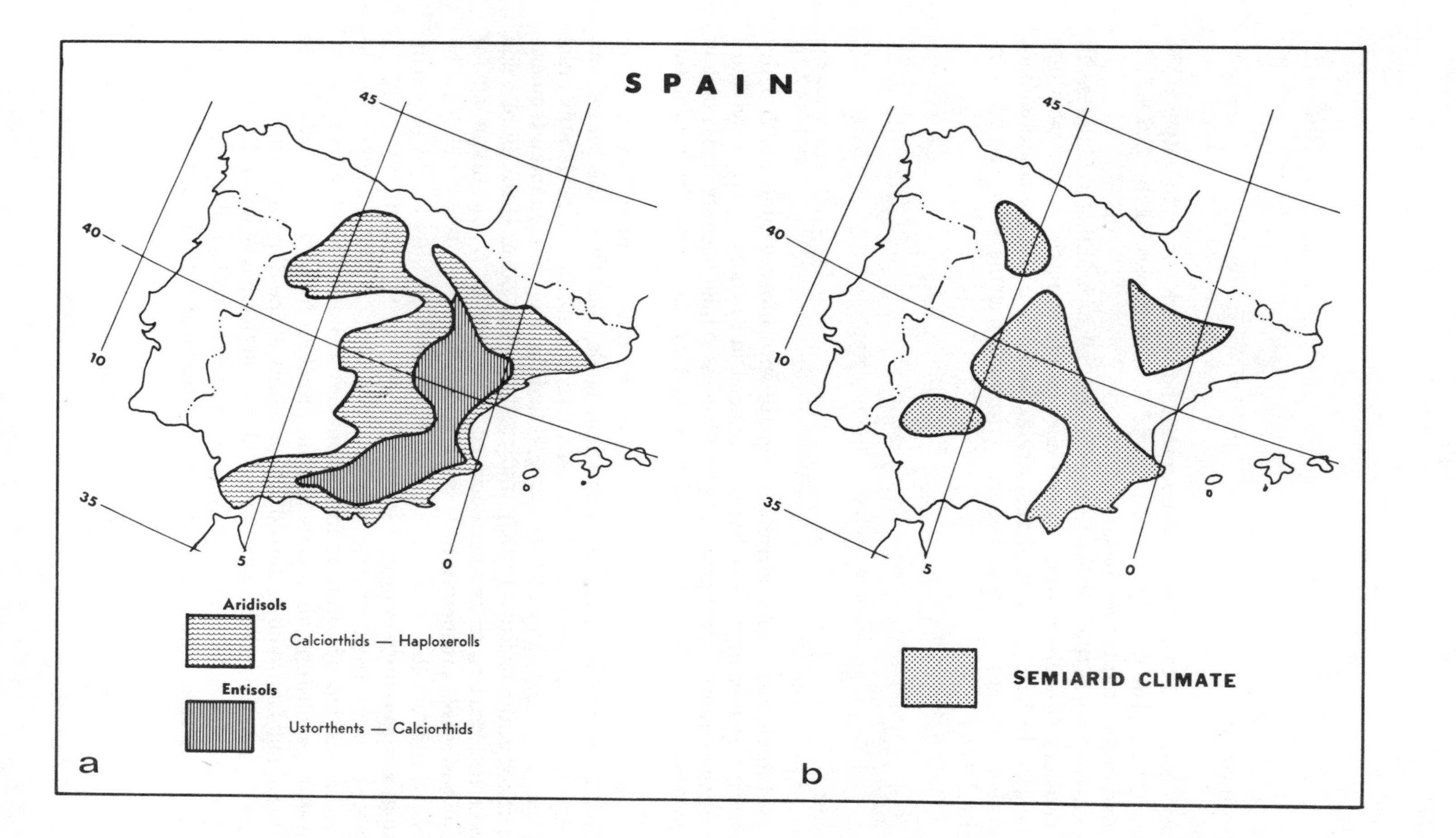

Fig.9. a. Arid-region soils of Spain. b. Arid regions of Spain (after Meigs, 1953).

TABLE 9.1

Climatological data for three weather stations in Spain

Station	Elevation (m)	Precipitation (mm)													Average annual temperature (°C)
		J	F	M	A	M	J	J	A	S	O	N	D	annual	
Burgos	861	38	38	53	48	61	43	20	18	36	51	56	51	513	10
Madrid	667	28	43	43	43	38	30	10	8	30	48	56	41	418	13
Almería	65	23	25	18	23	18	5	0	2	15	23	38	28	218	18

Source: U.S. Department of Commerce (1969).

generally are narrow but there are broad level areas along some of the major streams, particularly in the lower reaches of the Guadalquivir. Sand dunes are found along the Mediterranean coast in the south.

Vegetation

Man has had a pronounced effect on the vegetative cover of Spain. Little remains of the original vegetation, due to grazing, wood cutting, and cultivation. *Quercus* and *Thymus* are the dominant woodland types, with *Stipa* grasses found in some places. *Suaeda*, *Atriplex*, and *Caroxilon* occupy saline soils along the coast. On gypsiferous sites, *Gypsophila* and *Artemisia* shrubs are common and there is some *Agropyron* grass. *Pinus halepensis* is the principal tree of the mountains.

SOILS

Calcareous Aridisols and Entisols constitute the great majority of arid-region soils. Mollisols and gypsiferous soils are common, Salorthids are extensive in the lower Guadalquivir Valley, and saline soils present problems in several of the irrigated areas. Paleosols (buried soils) are said to be prevalent in much of the country.

Table 9.2 shows the area dominated by Aridisols and Entisols.

Aridisols

AR. Calciorthids — Haploxerolls

Aridisols and Mollisols are intermixed in this association, which includes the northern tablelands centered on Valladolid; the southern tablelands east,

TABLE 9.2

Area of soil orders of arid regions of Spain

Order	Area (km^2)	Percent of country
Aridisols	185,000	36.6
Entisols	69,000	13.7
Total	254,000	50.3

south, and southwest of Madrid; much of the Ebro Valley; the coastal zone; and uplands south of the Guadalquivir River. The topography is dominantly undulating to rolling, with some rounded mountains, steep slopes, and level flood plains. The native vegetation was an open *Quercus* woodland; at present about half of the land is in crop production, mainly cereals, and the remainder is largely grazing land for sheep and goats.

Calciorthids have a well-defined calcic horizon beginning at about 20—40 cm below the surface. Villar (1937) referred to calcic horizons as K horizons and noted the prevalence of such horizons throughout much of Spain. Many of the Calciorthids are gypsiferous and nearly all are calcareous to the surface. Surface-soil color varies from light brown to reddish brown to dark brown, most of the soils are moderately deep or deep, organic-matter content is low in the grazing-land soils but may be moderately high in some of the cultivated lands, and medium textures are the commonest. Salinity is a problem in some of the irrigated Calciorthids because of seepage from higher to lower areas. Erosion has brought the calcic horizon to the surface in many places.

Haploxerolls in this association may or may not have a cambic horizon below the mollic epipedon. They do have a calcic horizon in which the carbonate frequently is concentrated in what the Soviet soil scientists refer to as *white eyes.* They may be calcareous to the surface, medium textures predominate, organic-matter content usually is between 2 and 4% and base saturation is high. Soils are permeable and moderately deep to deep in the tablelands but may be shallow in the mountains. Subsoils may be stony.

Ustorthents, Gypsiorthids, Ustifluvents, Ustipsamments, and Salorthids are among the other great groups within this delineation.

Entisols

E. Ustorthents — Calciorthids

Shallow and moderately deep soils, many of them severely eroded, are found in the mountains and foothills of eastern Spain where the vegetation changes from woodland at the lower elevations to pine forest on the mountains. Irrigated narrow valleys are common and successful crop production is confined largely to the valleys. Ninety percent of the land is used for livestock grazing.

Ustorthents are shallow soils over generally unconsolidated rock. As with nearly all soils of arid Spain, the soils are calcareous and base saturation is high. Organic matter is variable with depth but usually is over 1% in the surface. Medium textures are dominant and slopes are steep. Stones and boulders are present in most of the profiles, on the surface as well as in the soil.

Calciorthids are calcareous to the surface, have a well-defined calcic horizon, occupy rolling to steep slopes, are shallow to moderately deep, and generally are gravelly or stony. Erosion is a major problem because of the heavy grazing pressure on the land.

Haplustalfs, Ustifluvents and Haplustolls are the principal associated great groups.

REFERENCES

Meigs, P., 1953. World distribution of arid and semi-arid homoclimates. In: UNESCO, *Reviews of Research on Arid Zone Hydrology. Arid Zone Res.*, I: 203—210.

Villar, H. del, 1937. *Soils of the Lusitano—Iberian Peninsula.* Murby, London (Translated from the Spanish by G.W. Robinson).

U.S. Department of Commerce, 1969. *Climates of the World.* Environmental Data Service, Environmental Science Services Administration, U.S. Department of Commerce, Washington, D.C., 28 pp.

Chapter 10

CHEMICAL PROPERTIES

INTRODUCTION

Soil is the product of the interactions among the five soil-forming factors: parent material (rock types), climate, vegetation, topography, and time (Jenny, 1941). In the arid regions, the major factors affecting the chemical properties of soils are climate, topography, and time, with vegetation (which is heavily dependent upon climate) and parent material playing secondary roles. When the average annual precipitation in both the hot and the cold arid regions is less than 250 mm, soils generally have neutral to alkaline reactions and a base saturation of 90% or more, contain less than 2% organic matter, have a calcareous horizon within the top 2 m of soil, are at least slightly saline, and have a clay mineral complex dominated by illite and montmorillonite. When annual precipitation exceeds 250 mm, differences appear which are related to the increased rainfall, variations in rainfall distribution patterns, position on slopes, and age of soil.

HORIZON DEVELOPMENT

Development of certain horizons in arid-region soils is an indication of the nature of the chemical environment in those soils. Subsurface horizons are of particular interest because they are influenced less than surface soils by transitory climatic events.

In the arid regions, uncultivated surface horizons (epipedons) fall into three categories: mollic, umbric, and ochric. The first two contain more organic matter and are darker than the last one; mollic and ochric epipedons generally are less acid than umbric epipedons; ochric (light-colored) epipedons are typical of soils of the drier part of the arid regions. Organic-matter content and soil reaction (acidity or alkalinity) are related to precipitation and temperature, mainly. As precipitation increases while temperature remains constant or as temperature decreases while precipitation remains constant, soil organic-matter content increases as the vegetation changes from shrubs to grassland. As the grasslands are replaced by savannahs and forests, organic matter decreases again, but not to the low level of the shrublands. The climate-related soil orders in that vegetation sequence are Aridisols (shrubs and desert grasslands), Mollisols (mid-to-tall grasslands), and Alfisols (savannahs and forests). Entisols, Inceptisols, and Vertisols will be interspersed among the other three orders. Typically, Aridisols and Alfisols have ochric and Mollisols have mollic epipedons.

A west—east transect along the Aridisol—Mollisol—Alfisol sequence of soils in the central United States shows a steadily decreasing base saturation and soil reaction as the climate becomes more humid. Soil nitrogen levels are related directly to organic-matter content. Aggregation increases as the weathering rate increases, from west to east across the Great Plains (Jenny, 1941). Argillic horizons are least common in Aridisols, commoner in Mollisols, and present in all Alfisols.

SUBSURFACE HORIZONS

The principal subsurface horizons representative of arid-region soils are calcic, petrocalcic, gypsic, petrogypsic, salic, and natric. Duripans and cambic horizons sometimes are significant. Formation of argillic horizons in arid climates has been a controversial subject for many years and probably will remain so for some time.

While it is customary to speak of calcic, gypsic, etc. horizons as subsurface horizons, they may be exposed on the surface if erosion has truncated the soil.

Calcic horizons are so useful as indicators of soil-forming processes that Marbut (1951) divided all the soils of the world into pedocals and pedalfers, at his highest level of soil classification. Pedocals had a calcium carbonate (cal) horizon; pedalfers had aluminum (Al) and iron (Fe) horizons in their subsoils but no carbonate accumulation. That classification was abandoned because of difficulties in applying it but the terms are still used, at times.

CALCIC HORIZONS

A calcic horizon is a subsurface horizon which contains appreciable accumulations of secondary calcium carbonate (and usually some magnesium carbonate as well). The calcic horizon ordinarily has more carbonate than the parent material (C horizon) or the underlying layers and is formed by the leaching of carbonates from surface soils or by the upward movement of carbonate-rich capillary water from a shallow groundwater table. In the U.S. Comprehensive Soil Classification System, well-defined criteria are used to identify calcic horizons. If the horizon is indurated, it is called a petrocalcic horizon.

Various terms have been used in the literature to refer to indurated and nonindurated carbonate horizons. Caliche is a term which is used widely — and ambiguously — in the western United States for carbonate horizons. It usually refers to an indurated horizon, but not always. Sometimes discrete hard carbonate nodules (concretions) or fragments of indurated carbonate horizons are referred to as caliche nodules or caliche fragments. Calcrete is rather widely used as a term for indurated calcium carbonate layers. Lime horizons are indurated or nonindurated calcareous horizons but a limestone horizon either contains indurated carbonate nodules or fragments or is a massive layer of indurated carbonate. One of the most ambiguous terms used in the soil science literature is that of "crusts". Carbonate crusts may be hard or soft, surface or subsurface carbonate layers; it is hazardous to assume that crusts are hardened surface features. Siliceous, gypsiferous, and saline crusts are as imprecisely defined as are carbonate crusts.

Depth to carbonate horizon

The relation between amount of rainfall and the depth to the calcareous horizon was examined by Jenny and Leonard (1934) in a transect across the Great Plains of the United States, from Colorado to Missouri. The transect was made along the 11°C isotherm and all samples were taken in upland loessial soils. Annual precipitation ranged from about 375 mm to nearly 1,000 mm. Fig.10.1 is a graph of the effect of precipitation on the average depth at which the horizon of carbonate concretions began. Surface soils were noncalcareous. In nonuniform parent material, the close relation between precipitation and depth of carbonate accumulation evident in Fig.10.1 would not be expected to be found. Using data from Marbut's *Soils of the United States* for soils developed on calcareous sediments in the arid western United States, Jenny (1941) noted that the depth to the calcic horizon varied erratically from soil to soil, probably due to differences in soil permeability, age, and parent material. However, in all cases the subsoil was more calcareous than the surface soil, which also was calcareous. On noncalcareous parent materials, depth to the calcic horizon varies in a similar erratic fashion but the surface soils usually are noncalcareous in the top few centimeters, at least. In practice, then, depth to the calcic horizon shows a high correlation with precipitation only when comparisons are made in soils of similar parent material and age.

Carbonate-horizon development

Carbonate-horizon development in gravelly and nongravelly sediments has been elucidated by Gile et al. (1966). Their investigation led to the establishment of four sequential stages of development (Table 10.1) to describe the nature and degree of carbonate accumulation leading to the formation of K horizons. In gravelly materials, carbonate accumulation begins as pebble coatings and progresses to thickened pebble coatings and filling of interpebble spaces, reaching the final stage when a hard laminar horizon forms above the plugged horizon. In nongravelly materials, carbonates accumulate as thin

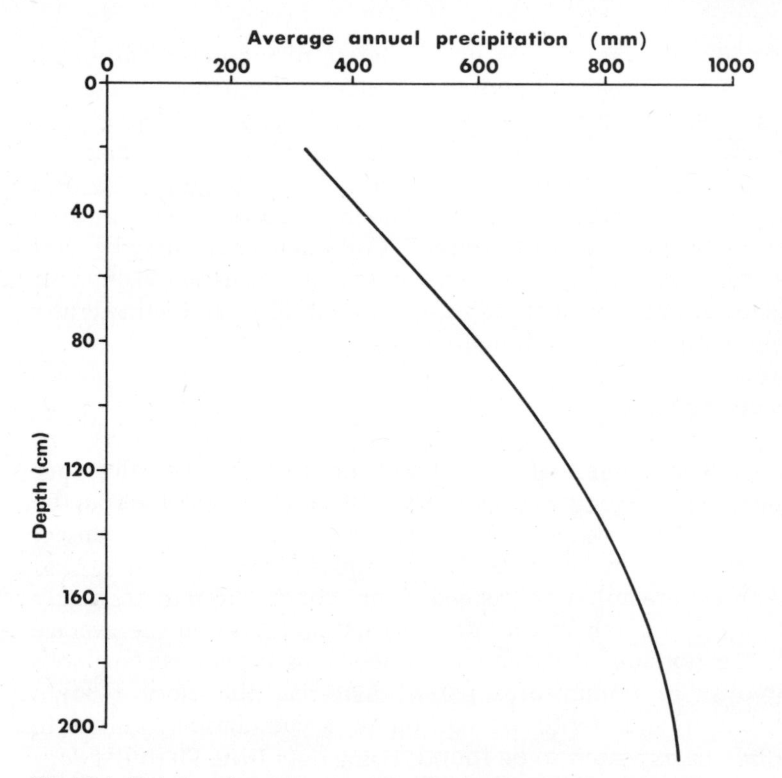

Fig.10.1. Relation between precipitation and depth to carbonate concretion horizon in loessial soils of the U.S. Great Plains (adapted from Jenny, 1941).

filaments and as faint coatings on sand grains. Later, carbonate nodules are formed and grow in size while the internodule spaces accumulate carbonates and ultimately become plugged. Finally, a laminar horizon is formed which — as in gravelly materials — grows upward as carbonate continues to accumulate. The laminar horizon typically has soft upper layers and hard lower layers, is several millimeters to a few centimeters thick, has a roughly horizontal orientation, is 70—95% carbonate, and has a white to pinkish color. In a vertical cut through a Stage IV soil, the surface horizon may or may not be calcareous, a calcareous (ca) horizon of soft and disseminated carbonate overlies the thin laminar horizon (soft on top, hard underneath) which, in turn, overlies a thick indurated plugged horizon; below the plugged horizon is a transitional horizon in which the degree of plugging decreases with depth. Petrocalcic horizons typical of Stage IV of carbonate accumulation develop earlier in coarse-textured sediments than in fine-textured sediments because plugging of pores occurs more rapidly in the former.

For a representative soil in southern New Mexico where the morphogenetic sequence of carbonate accumulation was diagnosed, radiocarbon measurements showed that the carbonate in the plugged horizon had an average age

TABLE 10.1

Stages of carbonate accumulation[1]

Stage	Diagnostic carbonate morphology		Youngest geomorphic surface on which stage of horizon occurs[2]	
	gravelly soils	non-gravelly soils	surface	age
I	thin pebble coatings	few filaments or faint coatings	Fillmore	$<$1,000 to 2,600 years
II	thick pebble coatings, some interpebble fillings	few to common nodules	Leasburg	$>$7,300 years latest Pleistocene
III	many interpebble fillings	many nodules and intermodular fillings	Picacho	Late Pleistocene
IV	laminar horizon overlying plugged horizon	(increasing carbonate impregnation)	Picacho	Late Pleistocene
	(thickened laminar and plugged horizons)		Jornada I	Middle Pleistocene
		laminar horizon overlying plugged horizon	La Mesa	Middle Pleistocene

[1]Adapted from Gile et al. (1966).
[2]At Las Cruces (New Mexico).

of 18,000 years, in the indurated laminar horizon an age of 14,000 years, and in the soft laminar horizon an age of 4,500 years. Carbonate in the calcareous horizon immediately above the soft laminar horizon was about 1,300 years old. Present-day rainfall in the area is between 200 and 250 mm. The middle of the plugged horizon occurs approximately at the average depth to which the soil is wetted by rain (50 cm for the above soil).

A hypothesis of indurated carbonate-horizon formation in layered soils has been proposed by Stuart and Dixon (1973). They theorized that carbonate and silica accumulation and ultimate cementation occurred at the interface where the soil texture changed abruptly. The textural change could be from fine to coarse or from coarse to fine. For accumulation to occur, water charged with dissolved carbonate and silica would need to penetrate to the interface, be held there, and then evaporate, leaving the dissolved substances to precipitate in the soil. Plants roots would assist in the deposition process by removing water through transpiration. The process described by Stuart and Dixon (1973) can account for the commonly observed cemented carbonate horizon lying immediately above a gravelly subsurface layer. It also accounts for lime horizons above bedrock. Subsequent steps in carbonate-horizon development presumably follow the stages proposed by Gile et al. (1966).

Carbonate sources

The source of the calcium carbonate could be: (1) the parent material of the soil; (2) deposition from the atmosphere, either as calcium carbonate particles or as other calcium salts which react with carbonic acid in the soil to form calcium carbonate; (3) upward movement of dissolved carbonates from a water table within a few meters of the soil surface; (4) mineralization of plant residues and the formation of calcium carbonate; and (5) deposition of carbonate from runoff or irrigation water. Magnesium is a common, but usually minor, constituent of soil carbonate complexes and it has its origin in the same way calcium does. Carbonate minerals consist principally of calcite, with some dolomite and lesser amounts of magnesite, aragonite, and siderite (Rozanov, 1951).

Worldwide, calcareous parent material and airborne calcium appear to be the major sources of soil carbonates. Dust traps in southern New Mexico captured calcareous dust each year they were in use (Gile et al., 1966) and rainfall in the area carries an average calcium content of about 3 mg/l (Junge and Werby, 1958), with the result that noncalcareous sediments ultimately became calcareous. Eriksson (1958), in his review of the contribution of sea salts to soil salinity, notes that loess in China, which is believed to be transported from arid Central Asia, is calcareous. He suggests that the source of the carbonate is atmospheric dust. Eriksson also notes the great distance — up to 1,000 km — cyclic salts are transported over the continents. Jackson (1957) mentions the possibility that carbonate in southern Australia soils is of aeolian origin.

Carbonate deposited by capillary water moving upward from a shallow water table probably is a significant source of soil carbonates in depressions. It also appears to account for the formation of the calcareous horizon in sandy soils on the fringe of the Kalahari Desert. Van der Merwe (1962) ascribes the development of the "limestone" horizon in Kalahari sand to solution of calcium in the underlying rocks by percolating rainwater and the subsequent upward movement and deposition of carbonate in a subsurface horizon when the sand dries. In Kalahari sand, carbonate nodules ranging up to the size of boulders constitute the "limestone", which is sufficiently porous to be easily penetrated by water and plant roots. The carbonate boulders have an extremely hard crust about 3 mm thick covering softer carbonate material inside the boulder. In situations where the carbonate source is from below, as in some places in the Kalahari Desert, the horizontally oriented laminar layer of Stage IV carbonate accumulation apparently does not form. The hard crust described by Van der Merwe as being found on the outside of the large boulder nodules may have some genetic similarity to a Stage IV laminar horizon. Durand (1959) believed that capillary movement upward from a carbonate-rich groundwater table accounted for the presence of indurated carbonate layers in the subsoils of low-lying plains and valleys in the Algerian Sahara.

Lobova (1960) cites several references in the U.S.S.R. literature which ascribe carbonate accumulation in desert soils derived from low-carbonate parent material to the mineralization of plant matter. According to this hypothesis, calcium and magnesium in plant tissues form carbonates during the decomposition process, leaving a residue of carbonates in the soil. As plants continue to grow and die, carbonates continue to accumulate in the surface, which, according to Lobova, accounts for the observation that carbonate levels in desert soils of the U.S.S.R. are highest in the surface horizon and decrease with depth. Rozanov (1951) attributed carbonate additions in Serozem soils to the mineralization of plant residues and to aeolian and alluvial deposition of carbonates.

Deposition of carbonates from runoff water accumulating in depressions and from irrigation water would be a slow process. The rate of deposition usually would be slow because of the low solubility of calcium and magnesium carbonates but the contribution could be significant if the geochemistry of the waters favored carbonate precipitation (Bower et al., 1965).

Due to the peculiarities of the U.S. Comprehensive Soil Classification System, the presence of a calcic horizon enters into the great group name only in the Aridisol and Mollisol orders and only in the absence of an argillic horizon. However, calcic horizons may be present in soils having argillic horizons, and they frequently are.

GYPSIC HORIZONS

A horizon in which the accumulated secondary gypsum (hydrated calcium sulfate) content exceeds, by defined amounts, that of the parent material or the underlying layer is a gypsic horizon in the U.S. Comprehensive Soil Classification System. Gypsum is a moderately soluble (2,000 ppm) salt; much more soluble than calcium carbonate but much less soluble than common salts such as sodium chloride and calcium chloride. Customarily, gypsum accumulates at greater depths in soils than does calcium carbonate because it is more readily leached. Significant amounts of gypsum are unlikely to be found in subsoils of well-drained, highly permeable soils unless the parent material was gypsiferous. Gypsum is common, however, in fine-textured soils of depressions where it occurs mainly as threads or soft nodules, and gypsum typically is present in or below the natric horizon of Solonetz soils (Natrargids).

Gypsum deposition

Few studies appear to have been made of the mechanisms by which gypsum accumulates in soils, despite its widespread occurrence in Aridisols. Hunt (1972) illustrated the effect of solubility on salt distribution in closed basins by using the analogy of evaporation of a solution of mixed salts in a dish. As evaporation proceeds, the least-soluble salts (carbonates) are precipitated first and form a ring around the top of the dish. With continued evaporation, the next most soluble salts (sulfates) precipitate, forming a lower ring, and finally the most soluble salts (chlorides) are precipitated in the bottom of the dish. The concentric ring pattern is typical of salt pans in Death Valley, California and elsewhere in the Great Basin and is associated with the drying of Pleistocene lakes. A similar vertical sequence of salts is noted in moderately or well drained soils under conditions of low rainfall: carbonates are concentrated in the upper part of the profile, gypsum below the carbonates, and the highly soluble salts at the greatest depth or leached completely out of the profile. In poorly drained soils with a shallow water table, the various salts tend to be mixed in the upper part of the profile.

Gypsum sources

Soil gypsum has its principal origin in gypsiferous parent material, airborne calcium sulfate, runoff water from gypsum-bearing rocks, and gypsum-containing groundwaters. Some soil calcium sulfate may be a consequence of the decomposition of plant residues but the amount probably is insignificant in comparison to the other sources. Crystalline gypsum is called selenite whereas noncrystalline fine-grained gypsum is called alabaster. Anhydrite is the name for anhydrous calcium sulfate.

Gypsiferous parent materials are particularly common in the southwestern United States, northern Mexico, Soviet Central Asia, the Middle East, and North Africa. Thick deposits of gypsum usually contain greater or lesser amounts of anhydrite and more soluble salts. Columnar beds of indurated gypsum have been described in the Soviet Union and Algeria, and they probably occur elsewhere, as well. According to Lobova (1960), the columnar beds of the Ust-Urt Plateau were formed by precipitation of gypsum from an ancient sea and are not related to present soil formation. On the other hand, Durand (1959) attributes the hard columnar gypsum in the Algerian Sahara to upward movement of capillary water from shallow water tables.

Airborne calcium sulfate could have its source in cyclic salts or in surface particles of salts (including gypsum) carried aloft by wind in the same manner calcium carbonate is. Junge and Werby (1958) found that both sources of sulfates contributed significantly to the sulfate levels in the air over the United States. They reported the lowest amounts of sulfate in rainwater over the oceans, the highest over the coastal regions, and intermediate levels in the interior. The presence of gypsum in well drained upland soils derived from nongypsiferous parent material points to airborne sources of the calcium sulfate; weathering is unlikely to provide the amounts of gypsum found in arid-region soils.

Runoff water from gypsum-bearing rocks is a rich source of soil gypsum. What is probably the largest surficial accumulation of nearly pure gypsum in the world — the White Sands of New Mexico — is a striking example of how important that source can be. Water running over and percolating through gypsum beds in the surrounding hills carries gypsum to a depression in the valley below. When the water evaporates, sand-size pure gypsum crystals are formed and blown out of the depression by strong winds, causing the development of gypsum dunes on the leeward side. Continued formation of new gypsum crystals in the depression and their subsequent removal by wind is responsible for a dazzling white 1,000 km^2 dune field. The Tularosa Valley, in which the White Sands lie, has gypsum in all of the valley soils. Runoff water probably accounts for much of the soil gypsum, but another important source is the gypsum in groundwaters. Gypsum beds of geologic, rather than pedologic, origin underlie the soils at various depths.

Less dramatic examples of the importance of runoff water to soil gypsum accumulations can be observed in the takyr-like playas and the salt pans, sebkhas, and chotts of arid-region closed basins. In the sodium-affected takyrs, as in Solonetz (Natrargid) soils, gypsum accumulates as veins in the subsoils, not in the surface. Gypsum mixed with more soluble salts may be found at any depth in the salt pans.

Upward movement of gypsum from shallow groundwater tables is a straightforward explanation for its presence in many soils underlain by gypsum beds. A comprehensive study by Durand (1959) in Sahara oases demonstrated to his satisfaction that hard gypsum (petrogypsic) horizons beneath cover sands were the result of the evaporation of gypsum-bearing groundwaters.

Gypsiorthids

Aridisols with gypsic horizons are Gypsiorthids. The presence of a gypsic horizon shows up in the great-group name only in the Aridisol order and only in the absence of an argillic horizon. In mapping soils, Gypsiorthids are not shown very frequently because most accumulations of gypsum are not great enough to qualify for a gypsic horizon. Many soils containing large amounts of gypsum are Calciorthids rather than Gypsiorthids. In general, all soils containing gypsum will be calcareous, but not all calcareous soils contain significant accumulations of gypsum.

SALIC HORIZONS

To be recognized as salic, a horizon must be at least 15 cm thick and have a secondary enrichment of salts more soluble than gypsum. It must contain at least 2% of soluble salts and the product of the horizon thickness (in centimeters) times percent salt must be at least 60. The soluble salts are chlorides of the common soil cations (calcium, magnesium, potassium, and sodium) and sulfates of the last three cations, plus carbonates and bicarbonates of sodium and potassium. Gypsum, plus calcium and magnesium carbonates and bicarbonates, also will be present in salic horizons, in most cases, and there may be minor amounts of other salts. From the standpoint of its adverse effect on plant growth, the most important soluble salt is sodium chloride.

Source of salts

The source of the more soluble soil salts is the same as for other salts: soil parent material, the atmosphere, runoff water, groundwater, and plant residues. By the very fact that they are soluble, salts comprising salic horizons can accumulate only under poor drainage or limited-rainfall conditions. Inadequate drainage accounts for salt accumulation in depressions, along sea coasts, and in irrigated lands. Salts can be present in well-drained upland soils if the parent material was saline and the climate arid.

Saline rocks exposed on the land surface and serving as the parent material for soils developed in place usually are of marine origin. Marine shales and gypsum or anhydrite rocks are the commonest saliferous soil parent materials. Saline soils developed from saline rocks are inextensive when compared to those in which the salts have accumulated through evaporation of runoff and groundwater.

A striking example of saline geologic material serving as the source of soil salts is the salt plugs of Iran (Dewan and Famouri, 1964). The exposed plugs standing several meters above the surrounding plain are salt domes in the Fars Series of geologic formations. Rainwater erodes the plugs and deposits large amounts of salt on the lower-lying soils.

Airborne salt distribution has been tabulated by Eriksson (1958), Junge and Werby (1958), and others. It is clear from Eriksson's review that cyclic salts and atmospheric dust contribute significant amounts of carbonates, sulfates, and chlorides to inland soils of all the continents. Downes (1961) cites research showing that the soluble salts in Australian soils have been brought in by air from the oceans. Northcote and Skene (1972) estimate that about 5% (386,000 km^2) of the soils of Australia are saline. In the absence of readily weatherable minerals, as in Australia, cyclic salts are the logical source for most of whatever salt accumulation occurs.

Salts in runoff water which accumulates in depressions and is evaporated are responsible for vast areas of saline soils in salt pans, playas, salinas, salares, sebkhas, and chotts. In some cases, the runoff water and groundwater are interconnected and both serve as salt sources. The largest salt desert in the world, the Dasht-i-Kavir in Iran, is an example of that phenomenon. More commonly, soils of topographic depressions are so slowly permeable that evaporation of runoff water serves as the primary source of the salt deposits.

In Australia, man's alteration of the vegetative cover has caused changes in the hydrologic regime that have led to the salinization of large areas of non-irrigated cultivated and pasture lands (Downes, 1961). The phenomenon has been observed in both semiarid and subhumid regions. Removal of deep-rooted trees and shrubs and their replacement with wheat and range grasses or by bare ground has raised water tables and increased seepage. According to Downes (1961), the main reason for the increase in seepage is that water use by wheat and pasture grasses (or bare ground) is less than that of the original trees and shrubs. Water that previously penetrated deeply into the soils and was transpired by the trees and shrubs no longer is taken up by the shallow-rooted plants. As a consequence, water tables have risen in some of the broad flat valleys, seepage has increased on the slopes, and subsoil salts have been brought to the surface. Halvorson and Black (1974) have described a somewhat similar situation in Montana where increased use of summer fallow has contributed to the expansion of saline seepage areas.

Saline groundwaters are a natural feature of many arid-region depressions but man bears the chief responsibility for their occurrence in irrigated lands. Overirrigation, canal seepage, and inadequate drainage installations lead to the development of high water tables, the assumulation of salts, and the ultimate abandonment of irrigated land. The Indus Plain in Pakistan is plagued with that problem even though the salinity of the Indus River and its tributaries is low. Other notable examples of the effect of inadequate drainage on soil productivity can be observed in the Imperial and Mexicali valleys of the United States and Mexico, the western Chaco in Argentina, the coastal valleys of Peru, and the Tigris—Euphrates Valley in Iraq. Soil salinity can truly be said to be the principal continuing problem in irrigated areas.

Decomposition of plant residues releases salts present in the plant tissues and, thereby, contributes to surface soil salinity. In most cases, the surface additions represent a redistribution of salts from subsoils to surface soil rather than an increase in the total amount of salt. Halophytes operate in this fashion to cycle salt from soil to roots to leaves and back to the soil. Plants contribute to soil salinity when their presence induces decomposition of relatively insoluble minerals and the formation of soluble salts. Leaf drop of *Atriplex confertifolia* was responsible for a large increase in the salinity of the soil under the plant as compared to that of the bare soil between the shrubs (Fireman and Hayward, 1952). Some species of *Atriplex* accumulate more salts in their leaves than do others (Wallace et al., 1973).

Salorthids

The only great group in which the presence of a salic horizon is apparent in the name is the Salorthid. Salic horizons are, however, found in other orders than Aridisols.

NATRIC HORIZONS

A natric horizon is an argillic horizon having prismatic or columnar structure and either an exchangeable sodium percentage (ESP) of 15 or more or a sodium adsorption ratio (an indirect measure of ESP) of 13 or more. In particular cases, exchangeable magnesium enters the definition of a natric horizon but the inclusion of magnesium is of doubtful validity. Natric horizons may or may not be saline.

Exchangeable (adsorbed) sodium is singled out for special attention because of the effect it has on the dispersion of clays in a low-salt soil. The criterion of 15% or more exchangeable sodium for a natric horizon follows from that proposed by the U.S. Salinity Laboratory Staff (1954) for an alkali (sodic) soil. When the ESP (Exchangeable Sodium Percentage) exceeds 15, a soil is said to contain sufficient exchangeable sodium to adversely affect plant growth and, by extrapolation to soil genesis processes, to significantly affect soil development. In Australia, Northcote and Skene (1972) consider that an ESP of 6 or more is enough to indicate that sodium salts have been an important genetic factor.

It is important to keep in mind that natric horizons must also be argillic horizons. Many soils in the arid regions have horizons with an ESP of more than 15, especially Vertisols, but the horizons are not natric unless they also are argillic. In fine-textured montmorillonitic soils such as Vertisols, water moves through so slowly that argillic horizons would develop only over very long periods of time even if there were no surface deposition or erosion or soil mixing.

Solonetz

Prior to the development of the U.S. Comprehensive System, soils with what are now called natric horizons were known as Solonetz. In a typical Solonetz, the A horizon has a neutral to alkaline reaction and is nonsaline, is of medium to moderately coarse texture, and abruptly overlies a columnar or prismatic fine textured, strongly alkaline, slowly permeable B horizon. The upper part of the B horizon is nonsaline, whereas the lower part may be saline and calcareous, and there may be veins or nodules of gypsum present.

Identification of Solonetz was made on the basis of morphological characteristics although more recently measurements or estimations of exchangeable-sodium percentage have been made and included in the U.S. definition. However, in the Soviet Union where the name originated the structure of the argillic horizon, not its ESP, determines whether a soil is a Solonetz. If the B horizon is an argillic horizon and has well-defined columnar or prismatic structure, the deduction is made that it must also have, or have had, a high content of exchangeable sodium. Similar reasoning has been used in other countries.

It frequently happens that chemical analyses of Solonetz B horizons show that the content of exchangeable magnesium is as great or greater than exchangeable sodium. That observation led to the conclusion that exchangeable magnesium played one of two roles. Either it was merely replacing the sodium which was responsible for the Solonetz structure, or it had similar effects on soil dispersion as did exchangeable sodium and, therefore, shared responsibility for Solonetz development. The former explanation appears to be the likely one. High levels of exchangeable magnesium can have direct adverse effects on plant nutrition, and those effects may have been confused with indirect soil effects.

In most cases, Solonetz have alkaline surface soils along with strongly alkaline subsoils. However, a widespread modification of Solonetz called solodized Solonetz has an acidic surface soil and a moderately alkaline subsoil. The reduction in alkalinity is presumed to be due to the formation of organic and inorganic acids in the surface soil and the gradual replacement of exchangeable sodium with hydrogen. The explanation is not quite that simple, as is obvious from the larger amount of exchangeable magnesium in solodized Solonetz than in Solonetz. Solodized Solonetz are widespread in the northern Great Plains of the United States, Canada, Australia, and the Soviet Union, in semiarid and subhumid regions. Within the drier sections of the arid regions, natric horizons usually are at least moderately saline and the A horizon will also be sodium-affected but probably not saline. Natric horizons are not common in the extremely arid climates.

Distribution

Worldwide, natric horizons are commonest in large and small depressions in the drier part of the arid regions and in both depressions and upland positions in the semiarid and subhumid regions. In order for them to form, there must be a source of sodium salts and there must be some leaching of the profile to remove excess salts, induce clay dispersion, and provide for clay eluviation. They are found in soils belonging to the Aridisol, Mollisol, and Alfisol orders and are most extensive in the Mollisol zone of the Soviet Union.

ARGILLIC HORIZONS

Over the years, arguments have arisen about whether or not an argillic (clay illuviation) horizon could develop in an arid-region soil. Early views in the Soviet Union, Australia, and the United States were that clay subsoils did not develop in dry climates (Jackson, 1957). Later, Nikiforoff (1937) claimed that clay formation did occur, in situ, in well-developed zonal soils of deserts but that clay eluviated from the surface horizon was unlikely to contribute significantly to the clay content of the subsoil. More recently, Smith and Buol (1968) and Gile and Grossman (1968) have found evidence that clay illuviation in subsoils does occur in the arid regions.

Horizon development

Argillic horizons are defined in the U.S. Comprehensive Soil Classification System as being illuvial horizons in which silicate clays have accumulated by downward movement from upper horizons. Clay formed in a subsoil by chemical weathering in situ does not cause an argillic horizon to develop although part or even most of the clay in such horizons does form in that manner; some fraction of the clay must be illuviated clay. When clay particles have been transported by water from an upper horizon to a lower horizon, they tend to be deposited with their long axes parallel to the ped (soil aggregate) surfaces on which they are deposited. Such oriented clay deposits are called clay skins. The presence of clay skins is, then, evidence of clay illuviation.

The resolution of the question whether argillic horizons do, in fact, develop in arid regions is complicated by numerous considerations. First, it is apparent that while some arid-region soils have argillic horizons, others occupying similar topographic positions do not. Second, argillic horizons are most obvious in soils occupying stable landscapes dating back to Late Pleistocene or earlier time. In that case, an argillic horizon could be a relict of earlier humid climates rather than a consequence of the present arid climate. Undoubtedly, many argillic horizons are relict horizons. Third, argillic horizons do not form on slopes or in depressions where the rate of erosion or deposition is equal to or greater than the rate of soil development. And fourth, soils developing in the high-carbonate sediments found extensively in the arid regions form argillic horizons at a much slower rate than soils developing in low-carbonate sediments (Gile and Hawley, 1972). The latter conclusion helps explain why some soils of a given age have argillic horizons while others of the same age and occupying the same topographic positions do not.

Effect of carbonate

Formation of argillic horizons in parent materials containing abundant fragments of calcareous rocks probably is impeded by the flocculating effect of calcium on clay particles. With flocculation, eluviation of silicate clay becomes more difficult since the floccules are relatively large and less subject to being carried in suspension through small soil pores. Also, clay formation is likely to be slower and clay adsorption less when soil minerals are coated with calcium carbonate. Knowledge of the carbonate content of the bedrock from which downslope alluvium was derived can be a useful guide to predictions of where Argids may be found.

Many older soils do not have argillic horizons even though their age, position, and parent material would indicate that they should have them. According to Gile and Hawley (1972), one reason may be that argillic horizons can be masked or obliterated by later carbonate accumulations which gradually engulf the argillic horizon. Calcium from the atmosphere would be the logical source of the calcium in the carbonate layer, in that case.

Clay skins

Clay skins are, at best, difficult to find in the more arid regions. The observation by Buol and Hole (1961) that clay skins can be destroyed by, among other things, soil mixing due to repeated drying and wetting led Smith and Buol (1968) to conclude that the near absence of clay skins did not mean that clay illuviation had not occurred. In their study of two soils in Arizona where the average annual rainfall was 277 and 417 mm, they found evidence from fine clay/coarse clay and calcium oxide/zirconium oxide ratios that clay had been transported from the A horizon to the B horizon even though only a few discontinuous clay skins were observed. In the area where the average annual rainfall was 417 mm, a low-carbonate soil on a 2—3% slope had an argillic horizon, whereas a high-carbonate soil on a 10—12% slope did not.

In another attempt to determine whether the presence of clay skins in a horizon of silicate clay accumulation was a valid requirement for argillic horizons, Gile and Grossman (1968) concluded that it was not in the 200—250 mm rainfall area in southern New Mexico which they studied. They contended that the presence of an argillic horizon was shown by the concentration of oriented clay particles on individual sand grains or pebbles and/or the presence of oriented clay within peds in the horizon of maximum clay content. Their study led them to believe that physical disturbance due to wetting and drying, adsorption of calcium carbonate, and root and faunal action disrupted or obliterated any clay skins which may have been formed.

Results of the two studies on clay skins implies that arid region soils, whether old or young, may not have well-defined clay skins even if the soils originally were developed under more humid climatic conditions where clay-skin formation could occur. Evidence that, on the contrary, clay skins do persist in dry climates is presented in Stace et al. (1968). Clay skins (called void argillans) identified as illuviated clay were detected in a Desert Loam in South Australia where the annual rainfall is 130 mm. Clay skins were common in the surface soil but became less common with increasing depth. There was no evidence of soil disturbance (pedoturbation) within the profile but the profile apparently had been truncated by wind erosion.

Returning to the question of whether argillic horizons can develop in arid climates, evidence bearing on the matter is ambiguous. Argillic horizons clearly are common in arid-region soils but they may be relicts of an earlier more humid climate. If the Gile and Grossman (1968) hypothesis concerning oriented clay particles is correct, then it is apparent from their studies that argillic horizons can form in arid climates in a relatively short time. They identified oriented clay (but not clay skins) in a soil more than 2,200 and less than 5,000 years old.

Soil color

One of the interesting sidelights to argillic horizon studies is the matter of soil color. Reddish-brown or red subsoil horizons in arid-region soils are good indicators of argillic horizons. Red colors are usually, if not always, due to the presence of free iron oxides on the surfaces of soil particles. As clay accumulates in subsoils, they gradually become redder and, consequently, redness becomes an expression of soil age. Illuviated clay is likely to be reddish brown rather than red because of colloidal organic matter adsorbed on the clay surfaces. The red iron sesquioxide color is the result of alternate wetting and drying of soils and the accumulation of oxides on soil surfaces. Iron oxidation is associated most closely with well-aerated soils in hot climates; thus, reddish colors are common in the hot arid regions. In cold arid regions such as those of the deserts of Mongolia, gray colors are dominant because iron-oxide accumulation is less rapid and higher organic-matter contents obscure whatever red colors may be present in the inorganic fraction. Calcium carbonate in coarse-textured soils tends to cause subsoils to assume a gray color as the carbonate content increases. In medium or fine textured soils possessing an argillic horizon, carbonates form nodules which — particularly in a reddish matrix — show up as prominent white spots.

OTHER HORIZONS

Of the other diagnostic subsurface horizons found in arid-region soils, duripans and cambic horizons are the ones of greatest interest. Neither cambic horizons nor duripans are restricted to arid-regions soils; cambic horizons are recognized at the great group level only in Aridisols (Camborthids) but they are found in soils of other orders.

Duripans

Silica cementation is presumed to be the principal cause of duripan formation. The degree of cementation varies from the strong cementation of very hard platy layers to the lesser cementation leading to the formation of massive clods. Duripan formation is believed to result from the leaching of soluble silica from the surface soil into the subsoil, followed by precipitation and dehydration of the silica during dry periods. Once the dehydration of silica has occurred, rehydration is difficult; duripans do not soften even after prolonged soaking in water or hydrochloric acid. In arid regions, silica accumulation is accompanied by calcium-carbonate accumulation and both act as cementing agents to hold soil particles together. Although silica is ubiquitous in soils, the richest source is the glass from volcanic-ash deposits. Consequently, duripans usually reach their maximum expression in areas where the soils have been affected by volcanic activity.

Silcrete

Silcrete consisting of as much as 95% silica is a prominent surface feature of central Australia (McKeague and Cline, 1963). When it forms a thick indurated cap on rocks, it is called *grey billy*; upon disintegration into stones or boulders, it is called *billy gibbers*. Gibbers make up the desert pavement of many stony deserts in Australia. While the mode of formation of silcrete is a subject of controversy, a likely explanation is that it represents a duripan or duripan-like layer which has been exposed through eroding away of the surface soil. There is some evidence that the silicified layer can form by dissolved silica moving upward from a shallow water table (Jackson, 1957).

Cambic horizon

Cambic horizons are weakly developed B horizons. Weak development shows up in arid-region soils in a redder color, a weak grade of structure, a small amount of silicate-clay accumulation, or evidence of carbonate movement into a lower horizon. Presumably, most cambic horizons will in time become argillic horizons if the soil is in a stable landscape position. Camborthids are not extensive because of the narrow range of characteristics that define them (Gile, 1966).

CLAY MINERALOGY

Clay minerals constitute the less than 2 μm size fraction of inorganic matter in soils. Particles less than 1 μm in diameter are said to be of colloidal size. In soils, colloidal organic compounds are present as well as clay minerals.

As a class, clay minerals come in a bewildering variety of forms and chemical composition. They include crystalline and noncrystalline silicates, oxides, hydrous oxides, and interstratified combinations of the crystalline silicates, plus practically all gradations from one to the others. Table 10.2 lists several of the clay minerals commonly found in soils. The expanding or nonexpanding character of crystalline silicates is not as well defined as the simple grouping indicates. Most chlorites, for example, are nonexpanding but some do have an expanding lattice.

Water-holding capacity and water permeability are two soil properties that are associated with clay content and mineral species. For all clay minerals, water-holding capacity and cation-exchange capacity vary directly with each other because both are dependent upon surface area. Hydrous oxides have the lowest cation-exchange capacity and water-holding capacity; montmorillonite is highest in both. Permeability to water, on the other hand, varies inversely with cation-exchange capacity except for allophane. Allophane is the exception in that it has a high cation-exchange and water-holding capacity but also is readily permeable. Swelling clays such as montmorillonite are responsible for the low permeability of Vertisols. Nonswelling kaolinite allows soils of the humid tropics to be highly permeable even though their clay content may be very high.

Weathering

According to Jackson (1964), the resistance of the clay minerals in Table 10.2 to weathering increases in the following order: illite—vermiculite—chlorite—montmorillonite—kaolinite—gibbsite—allophane—goethite. Illite is most easily weathered; goethite is most resistant to weathering. The greatest

TABLE 10.2

Classification of representative clay minerals

Mineral type	Mineral species	Characteristics	Cation-exchange capacity
Crystalline silicates	kaolinite	1:1 nonexpanding lattice	low
	montmorillonite	2:1 expanding lattice	high
	vermiculite	2:1 expanding lattice	high
	illite	2:1 nonexpanding lattice	moderate
	chlorite	2:2 nonexpanding lattice	moderate
	interstratified	mixture of silicates	variable
Noncrystalline silicates	allophane	amorphous	high
Hydrous oxides	gibbsite	aluminum hydrous oxide	very low
	goethite	iron hydrous oxide	very low

intensity of weathering occurs in hot wet climates, the least intensity in cold dry climates. In the humid tropics, minerals having a resistance to weathering equal to or exceeding that of kaolinite are the ones that dominate the clay-mineral complex. Only in arid regions would one expect to find large amounts of easily weathered minerals. The common primary minerals least resistant to weathering are gypsum and halite because of their moderate and high solubility, respectively, in water.

Considering its place in the weathering sequence, illite — if present in the soil parent material — should persist in arid climates and, at first glance, be expected to dominate the clay complex. However, since clay minerals more resistant to weathering than illite also would persist, the relative amount of any clay mineral in the soil would be dependent upon the amount in the parent material. Where clay formation occurs in arid-region soils, environmental conditions (alkalinity and soluble magnesium) favor the formation of montmorillonite except where soil potassium levels are high. The latter condition is favorable for illite formation (Marshall, 1964).

Occurrence

Whether due to occurrence in parent material or to formation in situ, illite is the commonest clay mineral in surface horizons and in weakly developed subsoils of upland arid-region soil (Buol, 1965). In depressions, montmorillonite is the dominant clay. Montmorillonite also is the dominant clay mineral in argillic horizons of North America and northern Asia; kaolinite and illite are present in lesser amounts. Illite, kaolin minerals, and interstratified clays are prevalent in surface soils and subsoils in Australia (Stace et al., 1968). Montmorillonite seldom is found anywhere in Australian arid regions except in depressions and Vertisols. Clay minerals in arid South African soils consist almost entirely of illite, montmorillonite, and mixed-layer (interstratified) clays (Van der Merwe and Heystek, 1955), and that same combination of clays seems to be dominant on the perimeter of the Sahara (D'Hoore, 1964).

Hydrous oxides of aluminum and iron have little genetic significance in arid regions although they are present in virtually all soils. Translocation of free iron and aluminum oxides would not be expected because neither the neutral to slightly alkaline soil reaction nor the well-oxidized soil atmosphere favor movement. Any movement that does occur is the result of the oxides being carried along on small silicate-clay particles.

SOIL REACTION

Soil reaction refers to the acidity or alkalinity of a soil. Acid soils are common in the humid regions where leaching of basic cations occurs; alkaline soils are widespread in the arid regions. Soil reaction is determined by the hydrogen-ion concentration of the soil solution and is expressed in terms of pH. The pH of a solution is the logarithm of the reciprocal of the hydrogen-ion concentration:

$$pH = \log [H^+]^{-1}$$

where hydrogen-ion concentration is given in equiv./l and commonly measured with a pH meter of the glass-electrode type.

Base saturation

In soils, the usual range in pH is from 4 (strongly acid) to 10 (strongly alkaline), with a pH of 7 representing neutrality. An indirect indication of soil pH is provided by base saturation, which is the percentage of the cation-exchange capacity occupied by basic cations (calcium, magnesium, potassium, and sodium, in the main). Base saturation is calculated from the following equation:

$$\text{Base saturation (\%)} = \frac{\text{basic cations (mequiv. g}^{-1}\ 10^{-2})}{\text{cation-exchange capacity (mequiv. g}^{-1}\ 10^{-2})} \times 100$$

Arid-region soils generally have base saturations of 80—100% due to the small amount of leaching in dry climates. When the base saturation is high, the hydrogen-ion concentration of the soil solution is low and the pH is high.

Relation to soil properties

Many properties of soils can be inferred, with some confidence, from pH. Soils with saturated paste pH's over 7.0 frequently contain considerable amounts of exchangeable (and available to plants) calcium, magnesium, and potassium. Available phosphorus, iron, and zinc are moderate to low, and microbial numbers are high if other environmental factors are favorable. At higher pH's, exchangeable sodium and sodium carbonate may be present in excess and cause undesirable physical and chemical conditions to develop in the soil. Corrosiveness of metals in contact with the soil is worse when soil pH is either high or low than when it is about neutral.

Effect of salinity and dilution

Unfortunately, soil pH is a function of, among other things, the salinity of the soil and the soil/water ratio used for determining pH. Dependence of pH on the soil/water ratio of the suspension in which it is measured frequently is ignored in reports, and pH data are given with no indication of the dilution factor used. Interpretation of such data is difficult or impossible.

The influence soil/water ratio and salinity can have on pH is illustrated in

Fig.10.2. At the high soil/water ratio represented by the saturated paste (approximately 1:0.4 for this soil), the pH varied from 7.1 in the highly saline soil to 8.0 in the nonsaline soil. Diluting the soil to the point where the soil/water ratio was 1:5 gave pH readings of 7.8 for the highly saline and 8.7 for the nonsaline soil. If the soil/water ratio had been greater (e.g., 1:10), the pH would have increased even more at any one level of soil salinity.

In either acid or alkaline soils, pH increases as soil dilution increases and it decreases as soil salinity increases. The difference between the pH of a saturated paste and a 1:5 dilution tends to be greater in sodic soils than in nonsodic soils. That observation has led to the suggestion that a difference of about 1 pH unit between the two readings indicates that the soil contains more than 15% exchangeable sodium. In many cases the relation holds reasonably well whereas in others it is quite unreliable. The reasons for the variability have not been determined but the nature of the clay minerals undoubtedly is of significance.

Highly saline soils have pH's around neutrality because that is the pH of the neutral salts constituting most of the solutes in the soil solution. In the presence of sodium carbonate, a strongly alkaline salt found in sodic soils, the pH of a saturated paste may be as high as 9.5 or 10. A nonsaline soil has a higher pH than a similar saline soil because basic exchangeable cations are able to form hydroxides in the soil solution, with a resultant increase in pH.

Removal of excess soluble salts during the reclamation of saline soils causes a rise in pH. The rise will be small if the dominant salts are chlorides and sulfates of calcium and magnesium (potassium seldom is a major constituent of soil salts). However, if the concentration of sodium salts is high

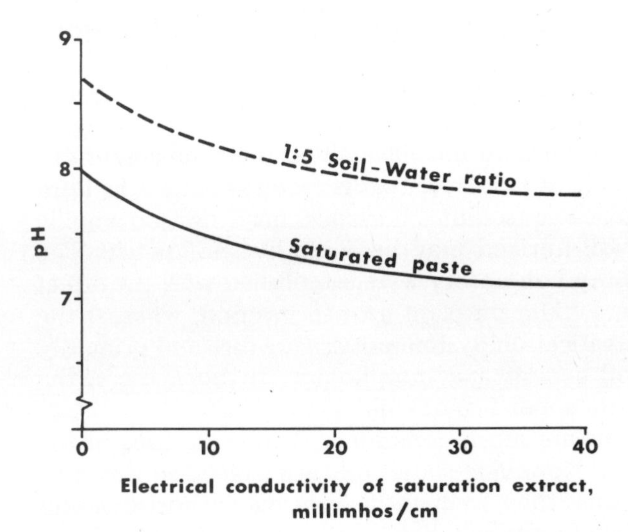

Fig.10.2. Relation of pH to soil salinity and dilution (adapted from Dregne and Mojallali, 1969).

enough to produce a sodic soil, rapid leaching of those salts will cause a large increase in pH due to sodium hydroxide and sodium-carbonate formation.

SALINE AND SODIC SOILS

According to the U.S. Salinity Laboratory Staff (1954), a saline soil has an electrical conductivity of the saturation extract (EC_e) of more than 4 mmhos/cm and an exchangeable sodium percentage (ESP) of less than 15. A sodic (alkali) soil has an ESP of more than 15 and an EC_e of less than 4 mmhos/cm. A saline—sodic soil has both an ESP greater than 15 and an EC_e greater than 4 mmhos/cm.

Electrical conductivity is an indirect measure of the amount of salt in solution. Conductivity figures in mmhos/cm ($EC \cdot 10^3$) can be converted to parts per million (ppm) of salt in solution by means of the equation:

$$\text{Parts per million} = EC \cdot 10^3 \times 640$$

The conversion factor of 640 is an approximate figure and is reasonably accurate up to a salinity equal to an $EC \cdot 10^3$ of 5(3,200 ppm). At higher salinities, the conversion factor is higher.

Another useful conversion is that changing the $EC \cdot 10^3$ of the saturation extract to the amount of soluble salt in the soil, expressed as ppm of the dry soil. The conversion formula is:

$$\text{ppm salt in soil} = \frac{EC \cdot 10^3 \text{ of saturation extract} \times 640 \times SP}{100}$$

where SP is the saturation percentage (percentage of water in soil at saturation).

Saline soils do not necessarily have salic horizons and sodic soils frequently do not have natric horizons. Many soils that qualify as saline according to the U.S. Salinity Laboratory criteria do not have either the total amount of salt or the concentration of salt in one layer that is required for a salic horizon. Similarly, high exchangeable sodium horizons may not be argillic horizons and, therefore, a soil horizon may be sodic but not natric. The criteria used by the U.S. Salinity Laboratory were established with the effect of salinity and exchangeable sodium on plant growth in mind, whereas the U.S. Comprehensive Soil Classification System criteria are directed primarily toward considerations of soil genesis and morphology. In both cases, the criteria apply to horizons in the upper 1 m of soil.

Saline and saline—sodic soils are much commoner than sodic soils in the drier parts of the arid regions. Nonsaline sodic soils are largely confined to semiarid and subhumid regions; they frequently have natric horizons and would be classified as Solonetz or solodized Solonetz in Soviet Union terminology.

Salt occurrence

The distinguishing characteristic of saline soils, from the agricultural standpoint, is that they contain enough soluble salt to affect adversely the growth of most crop plants. They are not, however, confined to agricultural soils nor to arid regions. Salt marshes, for example, are found along sea coasts in arid and humid climates. In general, saline soils have weakly developed profiles, saturated paste pH's near neutrality, and relatively good permeability to water. Adverse effects on plants are due to the higher osmotic pressure of the soil solution and the resulting increased resistance to water uptake, to impairment of metabolic processes, and to the toxicity of individual ions. Salts increase corrosion in uncoated steel pipes, and sulfates are particularly harmful to buried concrete. Structural stability of saline soils is moderate to high until they are leached and their clays become dispersed.

Ordinarily, soil salts are mixtures of a wide variety of soluble materials, among which chlorides and sulfates are dominant. In some cases, the cause of which is unknown, large amounts of calcium and magnesium chlorides accumulate. These chlorides are deliquescent salts that form organic complexes and impart a brown or black color to the soil. Because of the dark color of the surface soil, the affected areas have been called *black alkali* soils. In Iraq they are called *sabakh* soils (Buringh, 1960). They bear no similarity — except in color — to sodic soils also called *black alkali.* Field observations indicate that accumulations of calcium and magnesium chlorides occur mainly, if not entirely, through upward movement from shallow water tables, in irrigated and nonirrigated regions. Dark-colored chloride-affected soils have been noted in such different places as the Soviet Union, Iraq, Pakistan, Tunisia, Argentina, and the United States. Kovda and Dobrovolsky (1974) state that Soviet pedologists discovered an association in Azerbaijan and western Turkmenistan between soils salinized with sodium carbonate and calcium chloride and the presence of anticlinal oil and gas structures lying hundreds of meters below the surface. The Kura-Araks Valley southwest of the oil city of Baku in Azerbaijan has many "black alkali" chloride soils.

Salty layers can occur at any depth in a soil, depending upon such things as rainfall, depth to water table, and soil permeability. However, accumulations at or near the surface are typical of saline soils. When the salty layer is below 1 m, plants usually are not strongly affected by its presence. Reclamation techniques, therefore, are designed to keep the upper 1 m of soil relatively free of soluble salts.

Reclamation of saline soils

Leaching soils to remove soluble salts is the most effective method presently known to reclaim saline soils. The technique calls for the application of sufficient water to transport the salts below the root zone. For the procedure to be most effective, internal drainage must be adequate to maintain a water table — if one exists — at a depth of 2 m or more so that the salts will not rise again into the root zone. Leaching is inherently water-wasting, however necessary it may be to make the soil suitable for crop production, because water lost beneath the root zone does not contribute to the evapotranspiration requirements of a crop. Flushing water across a field to carry off surface salt deposits is an ineffective reclamation procedure.

Removal of salts by leaching reduces the salt hazard for plants but causes permeability to decrease and pH to increase. In saline—sodic soils, leaching can lead to a drastic drop in permeability and an increase in pH to the point where decomposition of roots occurs as the soil is changed from saline—sodic to sodic.

In principle, there is no need to leach salts out of the soil and into the drainage system. All that is required is the application of enough water to reduce the salinity of the root zone to tolerable levels, followed by irrigating frequently enough to maintain a downward hydraulic gradient to the bottom of the root zone. Under that water regime, salts can accumulate harmlessly at the edge of the root zone. Continued replacement by irrigation of water lost through evapotranspiration would insure that the hydraulic gradient was downward in the root zone and that salts would not move back into that zone.

Attractive though the theory is, it is beset by numerous practical difficulties. With conventional irrigation methods, only a well-designed and expertly operated sprinkler system can even approach the ideal, and then under only the best of conditions. Drip (trickle) irrigation is an unconventional system which does have the capability of reducing water applications to the evapotranspiration demand. Whether or not that capability can be utilized in field irrigation remains to be seen.

Boron

Boron is an element that deserves special attention in any discussion of saline soils. Sodium borate (borax) is the principal salt in which boron appears in soils. While boron is essential for plant growth, the range in concentration between a deficient and a toxic level in the soil is narrow for most crops. As little as 1 ppm of boron in the saturation extract can indicate that sensitive crops such as beans cannot be grown successfully.

Toxic amounts of boron are found occasionally in native soils but the principal source is irrigation water. Wells and streams of the southwestern United States, Alaska, Peru, Chile, and Australia, have been reported to carry potentially harmful quantities of boron. Very high boron concentrations have been measured in streams originating in volcanic areas of the Andes Mountains, with tributaries of the Rio Lluta in Chile containing as much as 40 ppm (Zambrano and Urrutia, 1961).

As with other soil salts, the only practical way to remove excess boron is by leaching the soil. Leaching is not as efficient in removing boron compounds as it is for chloride and sulfates. In one California field experiment, leaching was about three times as effective in reducing soluble salt levels as it was in reducing boron levels (Reeve et al., 1955).

Sodic-soil properties

Sodic soils exhibit the characteristics of Solonetz and other low-salt soils with natric horizons: strongly alkaline reaction in the horizon with the high ESP, dispersed clays, poor permeability of all except the most sandy soils, and at least small amounts of sodium carbonate. Their adverse effects on plant growth can be attributed to impaired aeration, restricted rooting depth, interference in nutrient uptake and plant metabolism, corrosion of root surface, and sodium toxicity. Due to the high degree of shrinking and swelling in sodium clays, sodic soils make poor base material for roads and airports and are unstable building sites.

Surface soils affected by exchangeable sodium have a characteristic polygonal cracking pattern when they are dry. Polygons average about 20—30 cm in diameter and typically have convex surfaces. The soil a few centimeters below the surface may be saturated with water at the same time the surface is dry and hard. Upon dehydration, cracks 1—2 cm across and several centimeters deep form, then close again when wetted. The cracks appear in the same place on the surface each time the soil dries unless it has been disturbed mechanically. If the soil contains significant amounts of organic matter, and is strongly alkaline, a dark film of dissolved or suspended organic matter appears on the surface. The dark color and high pH are the reasons why the soils are called *black alkali.*

Soil dispersion

Exchangeable sodium has long been known to cause dispersion of clays in a low-salt environment. Dispersion is weak in nonexpanding clays and strong in expanding clays, with the result that sodic soils containing only moderate amounts of expanding clay (e.g., montmorillonite) are slowly permeable to water and are unstable. Saline—sodic soils, on the other hand, contain sufficient soluble salt to maintain favorable permeability and stability conditions until they are leached and become nonsaline. Since clay dispersion facilitates eluviation of clays from the surface soil into the subsoil, argillic horizons are more likely to be developed in the presence of exchangeable sodium than in its absence. The Australian experience indicates that an ESP of more than 6 is enough to have a perceptible effect on the stability of soil aggregates (Emerson, 1967) and, by inference, on clay eluviation.

A controversy exists over whether soil cations other than sodium are dispersing agents. On the basis of their chemical properties, sodium and potassium should have similar effects in promoting dispersion of clays. Likewise, magnesium should act as a flocculating agent, in the same way calcium does. There is not general agreement, however, that such is the case. Some researchers contend that magnesium is a dispersing agent and potassium is not. Marshall (1949) reviewed the evidence on potassium and noted that the degree of swelling (dispersion) of potassium-saturated clays varied with the type of clay mineral. He suggested that swelling was low when the clays were of the potassium-fixing type (e.g., beidellite) and had been dried. In montmorillonite, which does not fix potassium, swelling was only about 23% less with potassium clays than sodium clays in an experiment by Baver and Winterkorn (1935). The zeta potential of potassium clays is slightly less than it is for sodium clays, indicating that dispersion values should be close for the two.

Magnesium has been implicated indirectly in the development of poor permeability, swelling, and dispersion in solodized Solonetz. No adequate explanation has been made of why the exchangeable magnesium is high in such soils and there has been no demonstration that magnesium has sodium-like effects on soils. Kelley (1951) says that the physical effect of exchangeable magnesium more closely approaches that of calcium than sodium.

Permeability

In a laboratory experiment on the effect of calcium, magnesium, sodium, potassium, and ammonium soils on permeability, Mojallali and Dregne (1969) treated Haplargid soil samples with decreasing concentrations of single cation chloride solutions. They found that when the end member (distilled water) of the sequence of solutions was applied, permeability went to zero in the sodium, potassium, and ammonium series. There was a gradual decrease in permeability in the sodium chloride series as salt concentration decreased but permeability went abruptly from high to zero at the end of the potassium and ammonium series. The soil was illitic but was not allowed to dry after treatment began. Calcium and magnesium soils had hydraulic conductivities of about 1 cm/hr when finally leached with distilled water. Martin and Richards (1959) measured large reductions in hydraulic conductivities with increasing amounts of exchangeable sodium, potassium, and ammonium in California soils that had been dried.

Reclamation of sodic soils

Reclamation of sodic soils is a two-step process. The first step involves the replacement of exchangeable sodium with calcium; the second step is to leach the resulting sodium salt from the soil.

Leaching, alone, of a calcareous soil on which a crop is growing can reclaim a sodic soil but the process usually is slow. Of the amendments used to bring about exchangeable sodium replacement with calcium, gypsum (hydrated calcium sulfate) is far and away the most-used material. It has the advantage of being inexpensive, generally available in the arid regions, nontoxic to plants, easy to handle, and moderately soluble. Highly soluble calcium chloride would be faster acting than gypsum but its use introduces an undesirable salinity problem and calls for more careful application techniques.

The chemical reaction of gypsum with a sodic soil is as follows:

$$CaSO_4 + 2NaX \rightleftharpoons CaX + Na_2SO_4$$

where X is the negatively charged exchange complex, consisting principally of clay. With leaching, the sodium sulfate is removed, the reaction goes to the right, and reclamation is accomplished.

Sulfur, sulfuric acid, and ferrous sulfate can be used effectively on calcareous sodic soils. Their application leads to the formation of calcium sulfate in the soil upon reaction with calcium carbonate. Sulfuric acid and ferrous sulfate produce calcium sulfate immediately or soon after application to a calcareous soil whereas sulfur must first be converted by microorganisms to sulfur dioxide, then to sulfuric acid, and finally to calcium sulfate. Microbial conversion of sulfur is slow.

Calculation of the amount of gypsum or other amendment to apply is done by analyzing the soil for exchangeable sodium and determining how much of the amendment is required to reduce the ESP to 10. An excess of about 25% above the calculated amount usually is recommended because the reaction is not 100% efficient.

PLANT NUTRIENTS

Nitrogen

Nitrogen is the plant nutrient most likely to be deficient in arid-region soils, followed by phosphorus, iron, and zinc. Crop responses to potassium have been noted in a few cases and sulfur deficiencies sometimes are found, especially in tropical Alfisols.

The nitrogen content of soils is almost entirely a function of the organic-matter content. The carbon/nitrogen ratio in grassland soils ranges from about 10 to 12; it is lower (8—9) in the more arid shrublands. Jenny and Leonard (1934), in their transect across the Great Plains of the United States along the 11°C annual isotherm, observed a pronounced positive correlation between soil nitrogen and annual precipitation. Jenny (1941) reported that organic-matter content showed the same trend. The soils were derived from loess and the original vegetation was grasses. Annual precipitation was between 375 and 1,000 mm.

Immediately after land has been put into cultivation for the first time, rapid microbial decomposition of readily decomposable organic matter begins and ample nitrogen usually is available for the initial crop. As cultivation continues without the return of crop residues or the growing of legumes, soil organic matter becomes more resistant to decomposition and nitrogen deficiencies appear. At that time, three alternatives are available to supply crop nitrogen: (1) allow the land to lie fallow to permit more organic matter to be decomposed before the next crop is planted; (2) grow a legume which can obtain its nitrogen from the atmosphere; or (3) apply nitrogen fertilizer (manure or commercial fertilizer). In semiarid dryland-farming areas, growing legumes in rotation with crops such as wheat is not a common practice, for various reasons, although some legumes (e.g., chick pea and pigeon pea) appear to have considerable potential as a nitrogen source under dryland conditions. Fallowing is the usual management technique for increasing the amount of available (inorganic) nitrogen, as well as moisture, in the soil. Applying commercial nitrogen fertilizer is risky in the drier part of the dryland-cropping areas because the added nitrogen will not increase yields if moisture happens to be much below average during the year in which nitrogen is applied. Much of the applied nitrogen will carry over to the next cropping year, however, and not be lost.

Under irrigated conditions and intensive crop-production practices, inorganic-nitrogen fertilizer applications are required nearly everywhere for maximum yields.

Phosphorus

Phosphorus availability to crops is low in practically all soils after they have been cropped for several years. In calcareous soils, native phosphorus generally is in the form of calcium phosphates of low solubility. Fertilizer phosphorus was believed to be changed rapidly to a form unavailable to plants in a calcareous soil but that has not been confirmed in long-time experiments (Schmehl and Romsdal, 1963). Since crops vary considerably in their requirements for added phosphorus, responses to phosphorus fertilizers are less certain than responses to nitrogen fertilizers.

Worldwide, there seems to be a positive correlation between redness of upland arid-region soils and phosphorus deficiency: the redder the soil, the greater the likelihood that it will be markedly deficient in phosphorus. How reliable that relation is remains to be seen. Phosphorus deficiencies are common in gray, brown and black soils; however, deficiencies appear to be more certain to occur in red upland soils. If this observation is correct, the reason may be that red soils are older and more highly weathered, contain less organic matter and calcium carbonate, and have more sites for phosphorus adsorption due to the iron-oxide coatings on soil particles.

Iron and zinc

Iron and zinc deficiencies are far less common than nitrogen and phosphorus deficiencies. They are most likely to occur in crops grown on calcareous soils and fertilized with phosphorus. Peaches, grain sorghum, and roses are among the plants most susceptible to iron deficiency whereas pecans, corn, and beans are quite sensitive to zinc levels in the soil. Soil deficiencies are due to reduced solubility of iron and zinc compounds in the presence of calcium carbonate and an alkaline pH. They are accentuated by phosphorus applications. Soil applications to correct deficiencies frequently are inefficient because of the adverse soil reactions.

Potassium and sulfur

Potassium and sulfur deficiencies in arid-region soils are of limited occurrence. Both are more likely to be found on weathered sandy soils than on fine-textured soils. Sulfur deficiency also is associated with soils derived from basalt in the United States and Australia. Customarily, irrigated soils receive sufficient sulfates in the irrigation water to meet crop needs although that is not always true. Sulfur deficiencies may be masked where ordinary superphosphate, which is about half gypsum, is used as the phosphorus fertilizer.

REFERENCES

Baver, L.D. and Winterkorn, H.F., 1935. Sorption of liquids by soil colloids. II. Surface behavior in the hydration of clays. *Soil Sci.*, 40: 403—419.

Bower, C.A., Wilcox, L. V., Akin, G.W. and Keyes, M.G., 1965. An index of the tendency of $CaCO_3$ to precipitate from irrigation water. *Soil Sci. Soc. Am., Proc.*, 29: 91—92.

Buol, S.W., 1965. Present soil-forming factors and processes in arid and semiarid regions. *Soil Sci.*, 99: 45—49.

Buol, S.W. and Hole, F.D., 1961. Clay skin genesis in Wisconsin soils. *Soil Sci. Soc. Am., Proc.*, 25: 377—379.

Buringh, P., 1960. *Soils and Soil Conditions in Iraq.* Directorate General of Agricultural Research and Projects, Ministry of Agriculture, Republic of Iraq, Baghdad, 322 pp.

Dewan, M.L. and Famouri, J., 1964. *The Soils of Iran.* FAO/UNESCO, Rome, 319 pp.

D'Hoore, J.L., 1964. *Soil Map of Africa. Scale 1 to 5,000,000. Explanatory Monograph and Map.* Commission for Technical Co-operation in Africa, Lagos, Joint Project No. 11, Publication No. 93: 205 pp. + 13 maps.

Downes, R.G., 1961. Soil salinity in non-irrigated arable and pastoral land as the result of imbalance of the hydrologic cycle. In: UNESCO, *Salinity Problems in the Arid Zones. Proceedings of the Teheran Symposium. Arid Zone Res.*, XIV: 105—110.

Dregne, H.E., 1969. Irrigation water quality and the leaching requirement. *N.M. Agric. Exp. Stn., Bull.*, No. 542.

Dregne, H.E. and Mojallali, H., 1969. Salt-fertilizer-specific ion interactions in soil. *N.M. Agric. Exp. Stn., Bull.*, No. 541.

Durand, J.H., 1959. *Les Sols Rouges et les Croutes en Algérie.* Direction de l'Hydraulique et de l'Equipment Rural, Service des Etudes Scientifiques, Alger, Etude Générale No. 7: 188 pp.
Emerson, W.W., 1967. A classification of soil aggregates based on their coherence in water. *Aust. J. Soil Res.*, 5: 47—57.
Eriksson, E., 1958. The chemical climate and saline soils in the arid zone. In: UNESCO, *Climatology, Reviews of Research. Arid Zone Res.*, X: 147—180.
Fireman, M. and Hayward, H.E., 1952. Indicator significance of some shrubs in the Escalante Desert, Utah. *Bot. Gaz.*, 114: 143—155.
Gile, L.H., 1966. Cambic and certain noncambic horizons in desert soils of southern New Mexico. *Soil Sci. Soc. Am., Proc.*, 30: 773—781.
Gile, L.H. and Grossman, R.B., 1968. Morphology of the argillic horizon in desert soils of southern New Mexico. *Soil Sci.*, 106: 6—15.
Gile, L.H. and Hawley, J.W., 1972. The prediction of soil occurrence in certain desert regions of the southwestern United States. *Soil Sci. Soc. Am., Proc.*, 36: 119—124.
Gile, L.H., Peterson, F.F. and Grossman, R.B., 1966. Morphological and genetic sequences of carbonate accumulation in desert soils. *Soil Sci.*, 101: 347—360.
Halvorson, A.D. and Black, A.L., 1974. Saline-seep development in dryland soils of northeastern Montana. *J. Soil Water Conserv.*, 29: 77—81.
Hunt, C.B., 1972. *Geology of Soils.* Freeman, San Francisco, Calif., 344 pp.
Jackson, E.A., 1957. Soil features in arid regions with particular reference to Australia. *J. Aust. Inst. Agric. Sci.*, 25: 196—208.
Jackson, M.L., 1964. Chemical composition of soils. In: F.E. Bear (Editor), *Chemistry of the Soil.* Reinhold, New York, N.Y., pp.71—141.
Jenny, H., 1941. *Factors of Soil Formation.* McGraw-Hill, New York, N.Y., 281 pp.
Jenny, H. and Leonard, C.D., 1934. Functional relationship between soil properties and rainfall. *Soil Sci.*, 38: 363—381.
Junge, C.E. and Werby, R.T., 1958. The concentration of chloride, sodium, potassium, calcium, and sulfate in rain water over the United States. *J. Meteorol.*, 15: 417—425.
Kelley, W.P., 1951. *Alkali Soils.* Reinhold, New York, N.Y., 176 pp.
Kovda, V.A. and Dobrovolsky, G.V., 1974. Soviet pedology to the 10th International Congress of Soil Science (the centenary of soil science in Russia). *Geoderma*, 12: 1—16.
Lobova, E.V., 1960. *Pochvy Pustynnoi Zony SSSR.* (*Soils of the Desert Zone of the USSR.* Issued in translation by the Israel Program for Scientific Translations, Jerusalem, 1967, 405 pp.; also cited as TT 67-51279.)
Marbut, C.F., 1951. *Soils: Their Genesis and Classification.* Soil Science Society of America, Madison, Wisc.
Marshall, C.E., 1949. *The Colloid Chemistry of the Silicate Minerals.* Academic Press, New York, N.Y., 195 pp.
Marshall, C.E., 1964. *The Physical Chemistry and Mineralogy of Soils. Volume I: Soil Materials.* Wiley, New York, N.Y.
Martin, J.P. and Richards, S.J., 1959. Influence of exchangeable hydrogen and calcium and of sodium, potassium, and ammonium at different hydrogen levels on certain physical properties of soils. *Soil Sci. Soc. Am., Proc.*, 23: 335—338.
McKeague, J.A. and Cline, M.G., 1963. Silica in soils. *Adv. Agron.*, 15: 339—396.
Mojallali, H. and Dregne, H.E., 1969. Relation of soil hydraulic conductivity to exchangeable cations and salinity. *N.M. Agric. Exp. Stn., Bull.*, No. 540.
Nikiforoff, C.C., 1937. General trends of the desert type of soil formation. *Soil Sci.*, 43: 105—131.
Northcote, K.H. and Skene, J.K.M., 1972. *Australian Soils with Saline and Sodic Properties.* Commonwealth Scientific and Industrial Research Organization, Melbourne, Vic., Soil Publication No. 27: 62 pp.

Reeve, R.C., Pillsbury, A.F. and Wilcox, L.V., 1955. Reclamation of a saline and high boron soil in the Coachella Valley of California. *Hilgardia*, 24: 69—91.

Rozanov, A.N., 1951. *Serozemy Srednei Azii.* (*Serozems of Central Asia.* Issued in translation by the Israel Program for Scientific Translations, Jerusalem, 1961, 550 pp.; also cited as OTS 60-21834).

Schmehl, W.R. and Romsdal, S.D., 1963. Materials and method of application of phosphate for alfalfa in Colorado. *Colo. Agric. Exp. Stn., Tech. Bull.*, No. 74.

Smith, B.R. and Buol, S.W., 1968. Genesis and relative weathering intensity studies in three semiarid soils. *Soil Sci. Soc. Am., Proc.*, 32: 261—265.

Stace, H.C.T., Hubble, G.D., Brewer, R., Northcote, K.H., Sleeman, J.R., Mulcahy, M.J. and Hallsworth, E.G., 1968. *A Handbook of Australian Soils.* Rellim Technical Publications, Glenside, S.A., 435 pp.

Stuart, D.M. and Dixon, R.M., 1973. Water movement and caliche formation in layered arid and semiarid soils. *Soil Sci. Soc. Am., Proc.*, 37: 323—324.

U.S. Salinity Laboratory Staff, 1954. *Diagnosis and Improvement of Saline and Alkali Soils.* U.S. Department of Agriculture, Washington, D.C., Agriculture Handbook No. 60: 160 pp.

Van der Merwe, C.R., 1962. *Soil Groups and Subgroups of South Africa. Repub. S. Afr., Dep. Agric. Tech. Serv., Sci. Bull.* No. 356: 355 pp.

Van der Merwe, C.R. and Heystek, H., 1955. Clay minerals of South African soil groups: III. Soils of the desert and adjoining semiarid regions. *Soil Sci.*, 80: 479—494.

Wallace, A., Romney, E.M. and Hale, V.O., 1973. Sodium relations in desert plants: I. Cation contents of some plant species from the Mojave and Great Basin deserts. *Soil Sci.*, 115: 284—287.

Zambrano, D.L. and Urrutia, A.B., 1961. *Calidad de las Aguas de Rio Lluta y afluentes.* Escuela de Agronomía, Universidad de Chile, Boletín Tecnico No. 9.

Chapter 11

PHYSICAL PROPERTIES

INTRODUCTION

Soil has been described as a three-phase system in dynamic equilibrium, the three phases being solid, liquid, and gas. The organic and inorganic solid phase is not merely an inert framework within which water and air are free to move independently of the solid matrix. Instead, interactions among the three phases occur. Infiltration of water into a dry soil illustrates some of the interactions. Infiltration is rapid initially (Fig.11.1), then slows as moisture films around the soil particles thicken and the clay minerals expand, reducing the size of the water conducting pores. Concurrently, biological activity increases and the soil air loses oxygen and gains carbon dioxide until anaerobic conditions develop, when additional gases are produced to further alter the composition of the soil atmosphere. If pressure is applied to the wet soil — by the weight of a tractor or a building — the soil mass can be deformed and long-lasting changes made in the physical, chemical, and biological environment. Drying sets in motion a different series of changes.

Water retention and movement command attention in any discussion of arid-region soils because of their obvious importance in plant growth and soil engineering. Directly or indirectly, water enters into the physical processes which shape the landscape, weather the minerals, form the soils, and determine the use to which the soils will be put. None of these processes is unique to the arid regions. Their impact, however, has special implications in moisture-deficient areas.

Aridity — and the conditions associated with it — contribute significantly to the intensity of wind and water erosion, the peculiarities of desert pavements, water repellency, soil permeability, the effectiveness of compaction of earth structures, piping, and other phenomena.

WIND EROSION

Movement of soil particles by wind begins when the pressure of the wind against dry surface-soil grains exceeds the force of gravity on the grains. Only when grains capable of being moved by saltation (jumping) are present in the soil can wind erosion occur. The first grain moved by the wind bounces against another grain and is propelled into the air where it gains horizontal momentum before falling back to the ground. When the grain strikes the

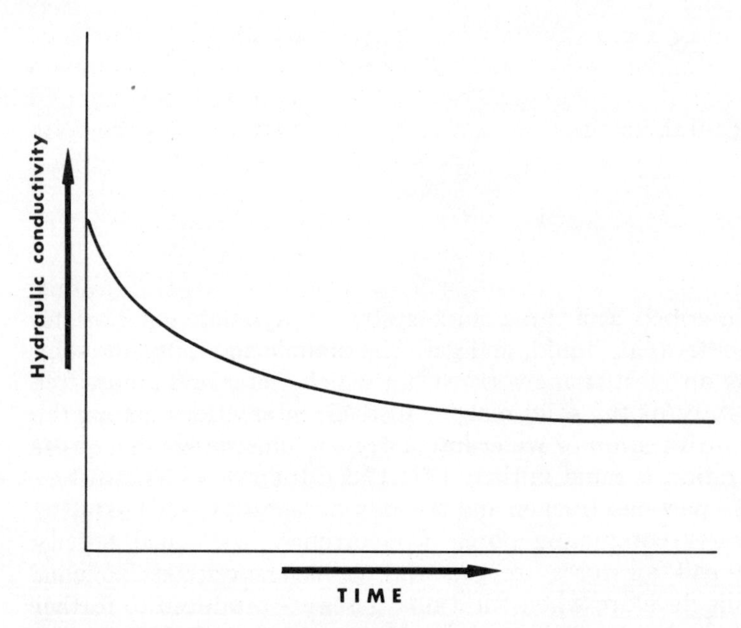

Fig.11.1. Generalized water infiltration curve for soils.

ground, it dislodges larger and smaller particles which continue the process. Proceeding downwind, the number of saltating grains increases to the maximum concentration the wind can carry, if the eroding area is large enough. Dust particles lifted from the ground by saltation may rise sufficiently high to be carried in suspension for long distances. Loess is predominantly silt-size dust particles picked up by the wind in dry Pleistocene valleys and deposited over huge areas in the United States, the Soviet Union, and China. Larger and heavier soil grains can be moved by surface creep, which consists of sliding or rolling along the surface rather than jumping. When erodible particles are present on a bare soil surface, soil movement begins when the wind velocity at 30 cm above the surface exceeds 20 km/hr.

The wind-velocity gradient above a cropped field is quite steep, going from zero wind velocity within the vegetative cover to nearly the maximum free-flow velocity at a height of 1 m above the crop. Wind movement strong enough to cause erosion is always turbulent flow, which enhances erosion by buffeting soil particles from many different directions in an erratic fashion. Wind velocity is accelerated on slopes and hilltops where windflow lines converge, thereby increasing the erosion hazard on knolls and crests of dunes.

Susceptibility of soil

In their comprehensive review of the physics of wind erosion, Chepil and Woodruff (1963) point out conditions which increase or decrease erodibility.

One of the most important is dryness, because only dry soil particles can be moved readily by wind. Moist or wet soils are nonerodible due to the cohesion of the water films surrounding soil particles. As farmers know, however, wind erodibility of sandy soils in the arid regions is greater after a light rain than before the rain. The reason is that the rain smooths the soil surface, forms a thin surface crust with loose sand particles on the surface, and permits a higher wind velocity near the surface. That combination of conditions contributes to easier initiation of saltation than does a rough dry surface.

Soil particles about 0.1 mm in diameter (fine sand size), whether individual grains or aggregates of grains, are most susceptible to wind erosion. Erodibility of particles larger than those passing through a 0.84-mm square-hole sieve is minimal; accordingly, that sieve size has been used in the United States to differentiate between erodible and nonerodible particles. Dust particles less than 0.02 mm are highly resistant to movement by wind until they have been disturbed in the saltation process.

Smooth soil surfaces, other things being equal, are more erodible than rough surfaces due to the higher wind velocities at the surface. Anything that interferes with the free flow of air across a soil surface helps to reduce wind erosion. In soils composed largely of erodible particles, roughening a dry soil surface may not be particularly effective in reducing wind erosion. Generally, however, surface roughening is effective. Sand accumulation in furrows demonstrates how useful tillage can be in holding soil on a field. Machines called *sand fighters*, designed for the specific purpose of turning up clods to roughen the soil surface, have been developed since Dust Bowl days and are in common use in the Great Plains.

Vegetation offers the best protection against wind erosion if the environment is favorable for plant growth. The least-desirable condition is a bare untreated soil surface. Destruction of the vegetative cover by cultivating marginal lands, overgrazing of livestock, and cutting trees for firewood is responsible for most of the man-made wind erosion. Where a vegetative cover is difficult to establish and maintain, as in the ergs and around oases in the Sahara, asphalt and other artificial covers may be the only materials which will stabilize erodible soils. Artificial covers, because of their expense and the problems associated with their maintenance, almost invariably are less desirable than living or dead vegetation.

Field arrangement

Wind erosion involves a process known as *soil avalanching* in which the number of particles in saltation increases downwind. Since the wind picks up more and more particles as it moves across an unprotected eroding field, width of field in the downwind direction influences the amount of erosion that will occur. Soil avalanching will be less on a narrow field than it will be on a wide field; strip cropping is based on that fact. Several narrow fields with wind barriers between them will be much less erodible than a single large field. Fields must be narrower if the soils are highly erodible than if they are not. Average recommended widths of crop strips on soils of different textures in the central Great Plains are given in Table 11.1. Granulated clay is erodible if the granules are less than 0.84 mm in diameter.

Along with size of field, the orientation of crop rows and barriers is an important consideration in wind-erosion control. Rows and barriers at right angles to the prevailing erosive-wind direction provide the greatest protection; in most cases, this means that the direction of erosive winds in spring and early summer should determine how rows and barriers are oriented.

Soil properties

Four soil properties are of particular importance in determining soil erodibility: texture, aggregation, organic-matter content, and amount and kind of calcium carbonate. As a first approximation, erodibility increases as sand content increases and it decreases as silt content increases. Clay has a variable effect which depends upon the amount of silt and sand present in the soil mixture (Chepil, 1955). Soils with 20—30% of expanding clays are the least erodible.

The quantity of water-stable particles less than 0.02 mm and more than 0.84 mm in diameter is correlated highly with resistance to wind erosion (Chepil, 1953). This is particularly true for the less than 0.02 mm fraction consisting of individual grains of silt and clay and of water-stable aggregates of silt, clay, and binding agents. The most-erodible soils, according to Chepil (1953), usually contain the greatest amount of water-stable particles between 0.05 and 0.42 mm in diameter. Water-stable particles form the matrix of the

TABLE 11.1

Average recommended widths of crop strips on soils of different textures in the central Great Plains

Soil Texture	Width of crop strips (m)
Sand	6
Loamy sand	8
Granulated clay	25
Sandy loam	30
Silty clay	45
Loam	75
Silt loam	85
Clay loam	105
Silty clay loam	130

Source: Woodruff et al. (1972).

nonwater-stable clods that are so important in determining soil erodibility. Clods are not as resistant to disintegration as water-stable aggregates but their presence is crucial in the control of wind erosion.

Somewhat surprisingly, soils with high organic-matter contents have been found to be more susceptible to wind erosion than low organic-matter soils. However, the effect is not quite so simple as that statement implies. Chepil (1955) added varying amounts of wheat straw and alfalfa to nine Great Plains soils and estimated erodibility from the proportion of dry clods more than 0.84 mm in diameter. His results showed that erodibility actually decreased during the first six months when the added plant residues were decomposing rapidly. A reversal of effect occurred after the period of rapid decomposition had passed, and tests made four years later showed clearly that erodibility was greatest on the soils receiving the largest amounts of residue. When the organic-matter content had stabilized, the erodible fraction had increased. Formation of new water-stable aggregates in the erodible size range and decreased mechanical stability of clods apparently are responsible for the increased erodibility.

The generalization about the adverse effect of high organic-matter levels on wind erodibility, then, applies only to stable (partially decomposed) organic matter. The highly erodible muck soils of Michigan are examples of how susceptible organic soils can be to wind erosion. However, fresh additions of plant residues help reduce erodibility. One of the major benefits of crop rotations which return organic matter to the soils is the regular addition of readily decomposable residues and the consequent increase in water-stable aggregates. Plant residues left on top of the soil are highly beneficial as long as they remain intact.

The fourth soil property of particular significance is the amount and kind of calcium carbonate (lime). Calcareous surface soils are notorious for their erodibility, and one of the hazards of levelling land for irrigation is the possibility of exposing calcareous subsoils. If the carbonate consists of a petrocalcic horizon, erodibility is lessened, but if it consists of powdery material, as it usually does, problems arise. Only in loamy sands and, presumably, sands does calcium carbonate decrease erodibility (Chepil, 1954). A reduction in the mechanical stability of clods appears to be responsible for the increased erodibility of calcareous soils (Chepil, 1954). Interestingly enough, the combination of added carbonate and organic matter (wheat straw) produced the highest degree of erodibility in the Chepil experiments with Great Plains soils.

WIND-EROSION CONTROL

A wind-erosion equation developed by the U.S. Department of Agriculture has proved to be useful in estimating potential wind erosion under existing conditions and in designing control measures (Woodruff et al., 1972). The equation is:

$$E = f(I', K', C', L', V)$$

where E is the average annual soil loss in tons per acre, I' is the soil-erodibility index (percentage of soil particles greater than 0.84 mm in diameter) and the percentage of slope, K' is soil-surface roughness, C' is the climatic factor (wind velocity and surface-soil moisture), L' is the width of the unsheltered field along the prevailing wind direction, and V is the vegetative cover.

Of the five factors influencing wind erosion, the climatic factor is the only one over which man has no control; soil-erodibility index, however, cannot be improved readily.

Permanent and temporary wind-erosion control utilizes one basic principle, that of providing surface barriers and cover to protect the erodible particles (Chepil and Woodruff, 1963). Any barrier that reduces surface wind velocity, be it fences, walls, trees, shrubs, grass, plant stubble, or crops, will reduce the erosion hazard. Similarly, anything that provides a noneroding cover for the erodible particles serves the same purpose. Wind-erosion control techniques are intended to provide one or both of these elements in an effective manner.

Minimum tillage

Numerous methods have been devised to control wind erosion. The most effective are those that conserve plant cover. For cultivated soils, stubble mulching and minimum tillage are two valuable methods for leaving protective plant residue on the soil surface. Stubble mulching refers to the maintenance of a plant residue cover throughout the year, whereas minimum tillage is a practice that conserves residues on the surface and also reduces the amount of soil pulverization due to tillage. Several pieces of tillage equipment have been developed for the specific purpose of leaving residues on the soil surface while land is being prepared for planting or weeds are being controlled. They include subsurface sweeps, rodweeders, offset disks, chisel plows, and other specialized machinery designed to meet particular needs.

Strip cropping

Strip cropping is a well-established method to reduce water and wind erosion. Alternate strips are planted to erosion-resistant crops (e.g., small grains, forage crops) and erosion-susceptible crops (e.g., cotton, peanuts, soybeans). Corn, sorghum, and millet are intermediate in resistance to erosion. For wind-erosion control, strips are planted in straight lines at right angles to the prevailing wind direction. For water-erosion control, the strips should follow the general contour of the land. Contour strips, though designed for water control, are of help in controlling wind erosion, as well.

In dryland-farming regions of the United States, strip width for wind-erosion control is usually the same for both erosion-resistant and erosion-susceptible crops. The choice of width is made on the basis of convenience rather than on the need for wide erosion-resistant strips. In irrigated fields where crop production is more intensive, one or two rows of small grains, corn, or sorghum may be grown to protect susceptible crops from damage by moving soil particles. Vegetable crops in the seedling stage are particularly susceptible to sand blasting as well as to excessive dehydration by hot dry winds.

Windbreaks

Windbreaks of trees have been planted extensively in the wheat-growing area of the Soviet Union for the dual purpose of reducing wind erosion and increasing snow accumulation. A major program of planting windbreaks in the Great Plains was undertaken in the 1930's in the United States but the popularity of windbreaks diminished in subsequent years. Problems arose with the adaptability of the trees and shrubs to arid conditions, the amount of land taken out of cultivation by windbreaks 10 to 12 rows wide, the long establishment period, and the interference with large-scale farm operations. Progress has been made on resolving the first two problems through plant-adaptation studies and the development of one to three row windbreaks.

Windbreaks consisting of masonry walls, snow fences, and similar structures can be effective although initial installation and maintenance can be costly and the downwind protected area is short. As a general guide, the downwind area protected by a wind barrier is considered to be equal to 10 times the height of the barrier. In practice, it varies from about 5 to 20 for most barriers. Low barriers need to be placed close together to protect the soil, which increases the installation cost and takes more land out of cultivation than if tall barriers are used. The higher the wind velocity, the shorter is the protected distance.

Emergency tillage

When other methods for wind erosion control have failed, reliance on emergency tillage becomes necessary on cultivated land. Emergency tillage consists of an attempt to create a rough, cloddy, surface by bringing subsoil clods to the surface. It is a temporary measure because the clods disintegrate rapidly.

Uncultivated land

On range lands where land values are relatively low, erosion control usually is limited by economics to restoring an adequate plant cover by controlling livestock grazing and by revegetation.

Protecting uncultivated land around populated areas and along highways and coastal beaches utilizes the same principles as for protection of cultivated land. Generally, however, the value of providing protection justifies greater expenditures to do so.

Among the more effective — and costly — measures are the application of stones and gravels, spraying asphalt or oil—latex mixtures on the soil, and spreading hay or straw and anchoring it to the surface. The latter two practices should be combined with the planting of adapted trees, shrubs, and grasses for continued protection. Spraying petroleum emulsions on sand dunes to provide temporary dune stabilization, followed by planting *Acacia* and *Eucalyptus* trees in holes is reported to have been successful in northern Libya where the rainfall is 300—400 mm (ALECSO/UNEP, 1975). Success also has been reported for partially embedding dead bushes in the sand and planting trees among the bushes. In both cases, the expense was justified by the need to protect nearby cultivated land and cities.

Another successful effort to stop destructive wind erosion has been reported for the Al Hasa Oasis in Saudi Arabia (Stevens, 1974; ALECSO/UNEP, 1975). Sand-dune control consisted of: (1) installing brushwood fences; (2) levelling dunes between the fences; (3) spreading soil from saline flats over the sands to give temporary stabilization; (4) planting *Tamarix aphylla* cuttings; and (5) irrigating the cuttings until they are well established. Annual precipitation is less than 100 mm. Sand-dune encroachment on the oasis is said to have been stopped and rather striking changes have occurred in the soil under the *Tamarix* during an eight-year period.

Control of wind erosion along highways in arid regions generally is accomplished by applying heavy petroleum products (asphalt, bitumen) and soil cements to sandy soils. Protection is restricted to a small strip along the highway, because of the expense, and does not prevent hazardous driving conditions from developing during major dust storms.

WATER EROSION

Water is, by all odds, the principal force shaping arid-region landscapes. Playas and arroyos are obvious examples of the results of water erosion although the newcomer who builds his home at the mouth of an arroyo apparently is not convinced that the dry bed would ever again carry a destructive flood of water, mud, and stones. Through the centuries down to modern times water erosion has caused devastation in former Roman grain fields of semiarid North Africa, the loessial hills of north China, dry northeastern Brazil, the Sahel, and in the western United States. The product of erosion is sediment, and deposition of soil has caused irrigation and flood-control reservoirs to lose capacity and effectiveness, filled irrigation canals, and damaged land and structures. Erosion control becomes increasingly difficult as the climate becomes more arid and vegetative cover sparser.

Initiation of erosion

Water and wind erosion have much in common because both water and wind movement obey the laws of fluid mechanics. In water erosion, soil movement is initiated by the pressure of moving water against soil particles or by the blasting effect of raindrops striking the particles. Raindrops exert a surprising amount of force when they hit the soil; splashed particles may rise as high as 1 m and cover a radius of nearly 2 m. Soil granules and clods can be shattered into smaller particles which reduce the soil infiltration rate and lead to more runoff and increased sheet erosion. Once the soil particles have been dislodged, they move by saltation, surface creep, and in suspension, just as they do in wind erosion.

Soil erodibility

As Wischmeier and Mannering (1969) have noted, soil erodibility is a function of interactions among its physical and chemical properties. There is no known simple measurement that will allow predictions to be made of a soil's susceptibility to water erosion under a variety of conditions. As a result, field tests of actual soil loss during and after rain storms have been required to provide erodibility data. Wischmeier and Mannering (1969) developed a rather complicated prediction equation based upon tests with a rain simulator on 55 widely differing soils from the north central United States, a humid region. The equation performed well in a comparison with values established previously on several other soils for which field erodibility was known.

In general, the most erodible soils are high in silt, low in clay, and low in organic matter. Susceptibility to water erosion seems to be most sensitive to changes in silt content; erodibility increases if organic matter is held constant. Very fine sand (0.05—0.10 mm in diameter) behaved more like silt than like sand. Increasing organic matter decreased erodibility, as did high levels of clay, with some important exceptions apparently related to organic matter—clay interactions. The amount of water-stable aggregates had only a small effect on explaining the variations in erodibility among the soils.

Silty soils

Loessial soils are the most silty soils of the arid regions, and their susceptibility to water erosion is well known. Silty layers are found in stratified alluvial soils but upland soils in which silt dominates the mineral fraction are confined mainly to loess and to soils derived from silty shales. The latter are highly susceptible to water erosion, especially if they are saline.

Erosion hazard

By and large, arid-region soils do not have textures that make them particularly susceptible to water erosion. Most upland soils are moderately coarse to coarse textured and many have a desert pavement that helps protect the surfaces from raindrop impact and reduce the rate of overland water flow. Fine-textured soils are concentrated in flat depressions where the erosion hazard is minimal even when soils are susceptible.

Nevertheless, water erosion causes problems everywhere in the arid zones except in deep sandy soils. A low amount of organic matter in the surface soil probably is the number one reason. The sparse vegetative cover exposes soils to the full force of raindrops striking clods and aggregates, and the paucity of organic matter means that resistance to disintegration is low. With dispersion of surface soils, infiltration is reduced drastically and runoff is high, which increases the erosive power of any given rain. Rain in the arid regions tends to come in high-intensity storms in which the rainfall rate greatly exceeds the infiltration rate (Keppel, 1960). The consequences of the high-intensity rain are rapid runoff and accelerated erosion.

Several other soil factors were included in the Wischmeier and Mannering study besides texture, organic matter, and water stable aggregates. Their identification improved the predictive value of the potential erodibility equation but the most important factors were texture and organic-matter content.

Additional factors

Water erosion is a function of a number of conditions other than soil erodibility. Among them are the steepness of slope, length of slope, amount of vegetative cover, the intensity and duration of rainfall, and cropping practices.

Soil loss increases about two and one-half times when the percent of slope doubles and about one and one-half times when the length of slope doubles. Longer slopes are more susceptible to erosion on the lower end because more water accumulates on long than on short slopes. Vegetation directly affects the erosion hazard in two ways: (1) plant canopies and residues reduce the impact of raindrops on the soil surface; and (2) anchored vegetation slows water movement across the land.

High-intensity rains present the greatest erosion hazard, for reasons mentioned earlier. Long-duration rains are more troublesome than those of short duration because infiltration rates decrease with time. The likelihood is greater that rainfall intensity will exceed the infiltration rate if the rain is of long duration.

Cropping practices have a strong influence on erodibility. If crop rows run up and down slopes, the erosion hazard is much greater than if the rows are oriented across the slope. Maintaining crop residues on the surface is preferable to burying them in the soil or to removing them. Leaving the soil bare between crops invites trouble on sloping land.

WATER-EROSION CONTROL

As in the case of wind erosion, a water-erosion equation has been developed for crop land by scientists of the U.S. Department of Agriculture and is described in a publication by Wischmeier and Smith (1965). The equation is:

$$A = RKLSCP$$

where A is the estimated annual soil loss in tons per acre, R is a rainfall factor dependent upon intensity and duration of rainfall, K is the soil erodibility, L is the length of slope factor, S is a slope-gradient factor, C is a cropping-management factor, and P is the erosion-control practice (e.g., strip cropping, terracing) factor.

The rainfall factor presently is beyond the control of man and only modest changes can be made — economically — in the soil erodibility factor. The others, however, can be altered radically by man's intervention and soil loss can be reduced to the three to five tons per acre considered tolerable by the U.S. Soil Conservation Service for agricultural land.

Improvement in the soil erodibility factor for cultivated land is dependent largely upon raising the organic-matter content by conserving crop residues and growing sod crops. Stabilizing soils for engineering purposes can be done by applying synthetic soil-binding agents to form large water-stable aggregates or by adding sand to change the texture.

Length of slope can be reduced by installing terraces or dams at intervals across the slope. Insofar as erodibility is concerned, effective slope length then becomes the distance between terraces or dams. Slope gradient can be altered by constructing terraces and partially or completely levelling the land between them. Bench terraces consisting of rock walls with soil between the walls have long been used to make possible production of crops on steep slopes in Africa, Asia, Europe, and South America. Protecting the terraces against breaks demands constant vigilance.

The cropping-management factor refers to the effect row crops, sod crops, crop rotations, the direction of rows, soil fertility, crop residue management, and similar decision matters have on erodibility. The least favorable cropping situation is one in which corn is grown in rows running up and down hill, the soil is infertile, no rotations are followed, and all crop residues are removed.

Erosion-control practices

Water-erosion control practices are based on slowing water runoff and holding the soil in place. The key item is the provision of barriers at right angles to the slope of the land. Barriers may be crop rows placed on the contour across slopes, strips of erosion-resistant crops planted on the contour, or terraces. Many kinds of terraces have been developed to meet specific needs. In the arid regions, terraces should be chosen to serve the dual purpose of controlling water erosion and conserving water.

A highly effective method to protect soil from the erosive power of water is to seal the surface over with asphalt or concrete. The net effect, however, may well be to increase destruction of land and buildings by increasing the volume and peak flow of runoff. Urban areas in the arid regions are particularly susceptible to that hazard because it further exacerbates the problem of high intensity rains.

For both wind and water erosion, establishment and maintenance of a good vegetative cover is the best way to control erosion. Destruction of vegetation by overgrazing, uncontrolled burning, and tree cutting is responsible for the impoverishment of vast areas in the Arab countries (ALECSO/UNEP, 1975) and elsewhere. The problem is becoming worse as numbers of livestock and people increase.

WATER MOVEMENT AND RETENTION

Hydraulic conductivity

Soil hydraulic conductivity (permeability) is a major factor in determining the use to which a soil can be put. Water erosion, reclamation of saline and sodic soils, stability of roads and building foundations, crop-production potential, and growth of trees, shrubs, and grasses are all affected to some degree by the ease or difficulty with which soils drain. Some of the factors determining hydraulic conductivity are the type of clay mineral, amount of water-stable aggregates, nature of the exchangeable cations, concentration of dissolved salts, organic-matter content, moisture content, hydraulic head, and thermal gradients. Soil water flow may be of the saturated or unsaturated type, depending upon whether pores are wholly or partially filled with water. Most of the time water flow is of the unsaturated type. Unsaturated hydraulic conductivity is frequently referred to as capillary conductivity.

The influence of soil moisture content, exchangeable cations, and soluble salts on hydraulic conductivity are particularly significant in arid region soils.

Soil moisture content

Saturated flow is the most rapid flow because the entire pore space is available to conduct water. As pores empty and flow is restricted to thinner and thinner films around the soil particles, conductivity decreases rapidly, as the soil moisture suction approaches 10 bar (Fig.11.2). In finer-textured soils than the sandy loam of Fig.11.2, conductivity generally does not decrease quite so rapidly.

A significant consequence of the low capillary conductivity of a dry soil is the development of what is called the "self-mulching" effect. Self-mulching refers to the condition following a rain where the surface soil dries and forms a natural mulch, thereby greatly reducing any further water loss by evaporation. That process, along with weed control, accounts for whatever water storage occurs during a fallow season in dryland farming areas. Water storage is greatest when the initial evaporation rate is high and a dry surface soil is formed rapidly (Fig.11.3). Sandy soils near Suez in Egypt are reported to be permanently moist below 50 cm (Abdel Rahman and El Hadidy, 1958), the moisture having been retained from years of heavy rainfall. Similar

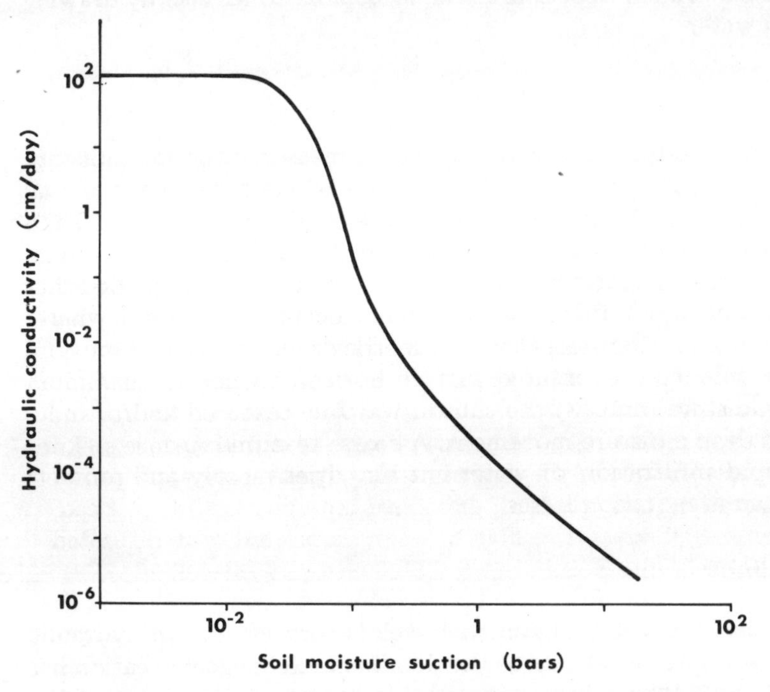

Fig.11.2. Hydraulic conductivity as a function of soil moisture suction (redrawn from Gardner, 1960).

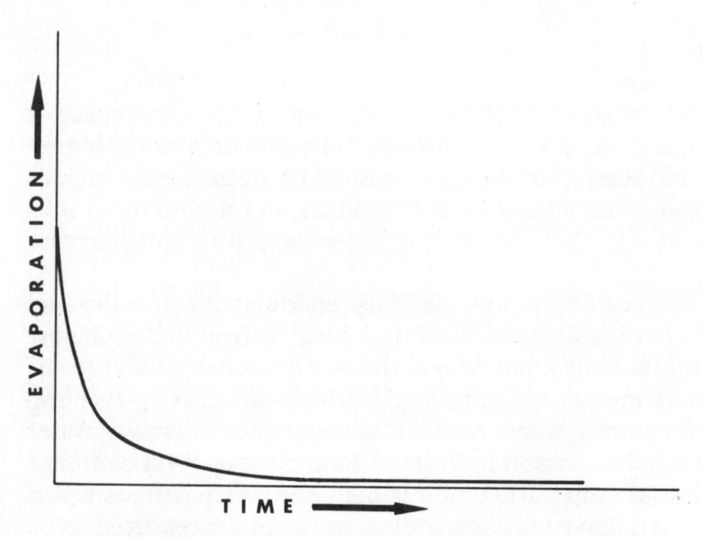

Fig.11.3. Evaporation from a bare soil during drying period.

observations have been made in barren sand dunes of North Africa where trees planted at 50—60 cm depths are said to be able to survive by drawing upon the stored water.

Water retention

In a ten-year field study, Herbel and Gile (1973) measured the soil moisture suction in several soils of southern New Mexico where the average annual precipitation is 220 mm. At the 25 cm depth, the tension was between 0 and 15 bar for an average of 40 days per year in a fine-textured Haplargid and an average of 212 days in a coarse-loamy Paleargid. The authors concluded that moisture conditions most favorable for plants occurred in areas where: (1) the landscape was level or nearly level, with little or no evidence of erosion; (2) there was a thin coarse-textured surface horizon to permit maximum infiltration of moisture; and (3) the subsoil was fine textured and/or indurated to prevent deep moisture movement. A coarse-textured surface soil not only permits rapid infiltration of water but also dries rapidly and protects subsoil water from evaporation losses.

Exchangeable cation

Exchangeable sodium and calcium have well-known effects on hydraulic conductivity. A soil having calcium as the dominant exchangeable cation has a higher conductivity than when exchangeable sodium is dominant, if the soluble-salt content is low. Calcium is a flocculating agent, sodium is a dispersing agent. Exchangeable magnesium resembles calcium in its effect on hydraulic conductivity and potassium — under certain conditions — is similar to sodium. These specific ion effects become operative, however, only when the electrolyte (soluble salt) concentration is low. At high electrolyte contents, hydraulic conductivity is relatively high irrespective of the nature of the salt. Electrolytes inhibit interlayer swelling of expansible soil minerals.

Electrolyte effects

Quirk and Schofield (1955) examined the effect of the four aforementioned cations, in the form of chloride salts at several electrolyte concentrations, on hydraulic conductivity. Their studies provided data on the degree to which hydraulic conductivity dropped as electrolyte content decreased in a soil saturated with exchangeable calcium, magnesium, sodium, or potassium (Fig.11.4). The initial concentrations of the four salts were: $CaCl_2$, 0.01 *M*; $MgCl_2$, 0.0316 *M*; NaCl, 1.0 *M*; KCl, 0.2 *M*. On this soil dominated by

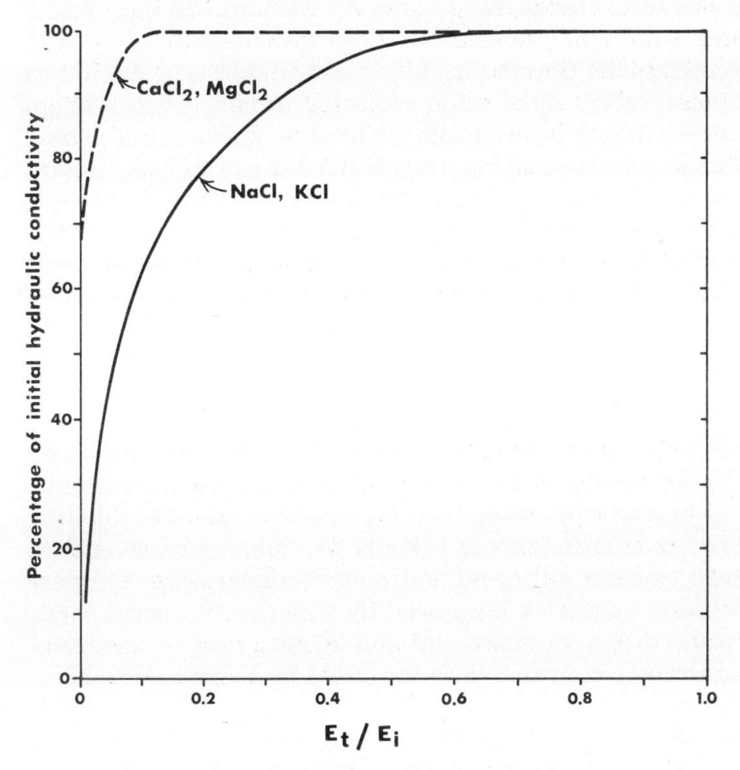

Fig.11.4. Hydraulic conductivity as influenced by type of cation and reduction in salinity of leaching water. E_i is the initial electrolyte concentration and E_t is the electrolyte concentration of the test solutions. Single-salt solutions were used. Figure adapted from data of Quirk and Schofield (1955).

kaolinite, illite, and vermiculite clay minerals, calcium and magnesium had similar effects on conductivity, as did sodium and potassium.

Field observations indicate that a little salt in irrigation water is better than no salt if soils are moderately or slowly permeable. At soil-solution salinities below about 100 ppm, maximum dispersion and swelling and minimum hydraulic conductivity occur. Beneficial effects on soil permeability attributed to applications of gypsum to slowly permeable, nonsaline, nonsodic soils probably are due to the increased flocculation caused by the gypsum rather than to any effect on exchangeable cations.

Reclamation with saline water

A common problem in the reclamation of saline—sodic soils is the occurrence of a sharp decrease in hydraulic conductivity as soon as the excess salts are leached out. When that happens, further reclamation of the sodic soil becomes very difficult due to inability to continue leaching in order to bring about the replacement of exchangeable sodium by calcium. An imaginative technique for coping with that problem has been described by Reeve and Doering (1966). It utilizes the flocculating effect of high-salt leaching waters to maintain high permeability rates while replacing sodium with calcium. Saline water of any cation composition can be used to increase soil permeability initially, after which the leaching water is diluted and its composition changed — if necessary — to make calcium the dominant cation. With that, replacement of sodium proceeds and reclamation can be accomplished. For the procedure to be effective, there must be an easily available source of salty water and a drainage system capable of handling large quantities of high-salt water.

Clay expansion

Hydraulic conductivity is influenced by a number of soil properties which are not unique to the arid regions. All other conditions being equal, hydraulic conductivity is lower in soils with expanding clay minerals (montmorillonite, vermiculite) than with nonexpanding clays (kaolinite, illite, hydrous oxides). Expanding clays swell to close soil pores and reduce conductivity. They are more subject to dispersion and may be moved by water to plug some pores, further reducing conductivity. Increased amounts of water-stable aggregates and organic matter operate to increase soil permeability by keeping pores open.

Temperature effects

Thermal gradients may be large in arid-region soils where surface temperatures rise to 58°C or more. At the same time, soil temperatures at 1 m depth are much lower. Water moves more readily when it is warm than when it is cold and, consequently, the water content of a warm surface soil tends to be lower than that of a cold surface. Since water vapor moves from a warmer to a colder region, soil moisture generally moves down in the vapor form during the day and up at night. The pronounced thermal gradients typical of arid-region soils make possible a simple solar distillation system devised by Jackson and Van Bavel (1965) for desert survival. The technique permits extraction of water from a relatively dry soil by condensing water on a plastic sheet.

LAYERED SOILS

Stratified soils containing fine and coarse textured layers are common in river flood plains of the arid regions. When a coarse sand or gravel abruptly underlies a medium to fine textured surface soil, an interesting effect on moisture retention occurs. Instead of water moving readily into a dry coarse-textured layer following rain or an irrigation, it is temporarily restrained until the soil moisture suction above the coarse layer approaches zero. At that time, water enters the coarse layer and moves down the profile. If no more water is added, movement ceases fairly soon and drainage stops as the hydraulic conductivity of the coarse layer becomes negligible.

A consequence of the layering effect is an increase in moisture content of the surface layer above what it would be if the soil were not layered. Miller and Bunger (1963) measured a 50—60% greater moisture content in a sandy loam overlying sand or gravel than in similar unlayered soils. If there is a gradual transition from a finer-textured surface soil to a coarse-textured subsoil, water movement is not restricted and moisture contents attain values typical of the corresponding uniform soils.

Two effects, one favorable and one unfavorable, can be attributed to the presence of a coarse layer within 30—50 cm of the surface. The favorable effect is a greater storage of water in the root zone above the coarse layer and less loss of water by deep drainage. The unfavorable effect is the stopping of root growth at the coarse layer, presumably due to a temporary saturation of the soil at the interface between the layers and the inability of the roots of some plants to survive in the saturated zone. The latter phenomenon has been called "root pruning" because the roots at the interface give the appearance of being pruned at that point.

Indurated carbonate formation

Formation of indurated calcareous layers immediately above a coarse layer has been described by Stuart and Dixon (1973). They associated the accumulation of silica, carbonates, and other salts with the greater moisture retention above the interface and the deposition of the dissolved constituents at that point as water is removed by evapotranspiration. Their explanation of the reason for carbonate accumulation above the coarse layer is similar to that described earlier for carbonate accumulation on top of bedrock.

WATER-REPELLENT SOILS

Runoff from many arid-region soils is accelerated, temporarily, by the water-repellent character of a film surrounding individual particles. The hydrophobic film consists of organic materials of unknown composition. Water repellence in low organic matter soils is a temporary condition where the soil repels water for as much as 45 sec before wetting occurs. Once wetting has occurred, water moves through the soil in the usual manner. Repellence is reestablished when the soil becomes thoroughly dry again.

Burning shrublands in semiarid regions has been observed to markedly increase water-repellence, the effect varying with the heat of the fire (Scholl, 1975). The practice of burning savannahs in Africa may contribute to increased water-repellence and to greater water erosion damage from subsequent rains.

PIPING

Piping is a type of water erosion in which the subsoil is removed laterally from under the surface soil. Pipe erosion is also known as tunnel erosion. After piping has occurred, the surface soil frequently collapses into the tunnels. Entire sections of soil several meters in diameter have been known to drop vertically — and in one piece — to a depth of 3 or 4 m. Piping is most likely to occur in soils that contain large amounts of gypsum and other soluble salts, are high in exchangeable sodium, or have an impermeable subsoil layer beneath an unconsolidated surface soil.

According to Fletcher et al. (1954), the conditions required in order for piping to occur are: (1) the surface infiltration rate must exceed the permeability of a subsoil layer; (2) there must be an erodible layer above the water-retarding layer; (3) the water above the retarding layer must be under an hydraulic gradient; and (4) there must be an outlet for the lateral flow.

Loose gypsiferous soils are particularly subject to piping. Water dissolves the salts, leaving voids and tunnels into which more water flows until the tunnels become large and troublesome. Frequently, the tunnels will collapse under the weight of the soil or when heavy machinery passes over. Control of the piping problem consists of moistening and compacting loose surface soils, growing plants having extensive root systems that will stabilize the soil, and avoiding accumulation of water in depressions.

DESERT PAVEMENT

Broad alluvial fans, piedmont slopes, and plateaus in the arid regions often are covered with rounded and angular stones which form the well-known desert pavement. The underlying soil frequently is gravelly and stony but may be medium or fine textured. Pavements are sometimes called desert armor because they help protect the soil from wind and water erosion. They make good travelling surfaces for vehicles when the stones are small. Hamadas and serirs of the Middle East and North Africa, as well as the gibber plains of Australia, are plateaus covered with rocks and boulders. Regs are gravel-covered fans and piedmont plains that occupy vast areas in the Sahara. Many desert pavement stones are covered on their upper side with a dark glossy film called *desert varnish.* The color is due to precipitation of iron and manganese oxides on the rock surface.

Considerable controversy exists on the origin of the smaller-sized rock fragments and gravels. They are variously believed to have been: (1) deposited by running water; (2) left behind when wind and water erosion removed finer material; or (3) brought to the top by alternate wetting and drying of the soil. The latter explanation is invoked when a desert pavement overlies a nearly stone-free soil. Deposition by water certainly occurs but it probably is of limited importance over large areas. Wind and water erosion appear to account for most of the desert pavements, with upward movement of stones from a shallow depth being significant when the soil is medium to fine textured. Desert pavements typically are found in areas where plant cover is sparse and there is little impediment to wind and water erosion.

Beneath the desert pavement may be a vesicular (porous) layer 3—10 cm thick (Springer, 1958). Vesicular layers are not restricted to soils having desert pavements, and they may be less than 3 cm thick. Vesicle formation is attributed to entrapment of air in dry, dusty soils during a rain. As the entrapped air escapes, it leaves behind voids which give a puffy appearance to the soil.

GILGAI

Many Vertisols exhibit a type of microrelief known as *gilgai.* It consists of a series of mounds varying from a few centimeters to a meter or more above the intervening depressions. Gilgai relief is confined to heavy clay soils subject to deep cracking. Cracks formed during the dry season are filled with loose soil falling in from the surface. Upon wetting, the clays expand and part of the soil is forced upward. In time, mounds and depressions are formed, sometimes in lines across the slope, other times in a repeating pattern of roughly circular hummocks and hollows.

REFERENCES

Abdel Rahman, A.A. and El Hadidy, E.M., 1958. Observations on the water output of the desert vegetation along Suez Road. *Egypt. J. Bot.*, 1: 19—38.

ALECSO/UNEP, 1975. *Re-greening of Arab Deserts.* Arab League Educational, Cultural and Scientific Organization and United Nations Environment Programme, Cairo, UNEP Project No. 0206-74-002, Final Report.

Chepil, W.S., 1953. Factors that influence clod structure and erodibility of soil by wind: II. Water-stable structure. *Soil Sci.*, 76: 389—399.

Chepil, W.S., 1954. Factors that influence clod structure and erodibility of soil by wind: III. Calcium carbonate and decomposed organic matter. *Soil Sci.*, 77: 473—480.

Chepil, W.S., 1955. Factors that influence clod structure and erodibility of soil by wind: IV. Sand, silt, and clay. *Soil Sci.*, 80: 155—162.

Chepil, W.S. and Woodruff, N.P., 1963. The physics of wind erosion and its control. *Adv. Agron.*, 15: 211—302.

Fletcher, J.E., Harris, K., Peterson, H.B. and Chandler, V.N., 1954. Piping. *Am. Geophys. Union, Trans.*, 35: 258—263.

Gardner, W.R., 1960. Soil water relations in arid and semi-arid conditions. In: UNESCO, *Reviews of Research on Plant—Water Relationships in Arid and Semi-Arid Conditions. Arid Zone Res.*, XV: 37—61.

Herbel, C.H. and Gile, L.H., 1973. Field moisture regimes and morphology of some arid-land soils in New Mexico. In: *Field Soil Water Regimes.* Soil Science Society of America, Madison, Wisc., pp.119—152.

Jackson, R.D. and Van Bavel, C.H.M., 1965. Solar distillation of water from soils and plant materials: a simple desert survival technique. *Science*, 149: 1377—1379.

Keppel, R.V., 1960. Water yield from southwestern grassland. In: B.H. Warnock and J.L. Gardner (Editors), *Water Yield in Relation to Environment in the Southwestern United States.* Sul Ross State College, Alpine, Texas, 74 pp.

Miller, D.E. and Bunger, W.C., 1963. Moisture retention by soil with coarse layers in the profile. *Soil Sci. Soc. Am., Proc.*, 27: 586—589.

Quirk, J.P. and Schofield, R.K., 1955. The effect of electrolyte concentration on soil permeability. *J. Soil Sci.*, 6: 163—178.

Reeve, R.C. and Doering, E.J., 1966. The high-salt-water dilution method for reclaiming sodic soils. *Soil Sci. Soc. Am., Proc.*, 30: 498—504.

Scholl, D.G., 1975. Soil wettability and fire in Arizona chapparal. *Soil Sci. Soc. Am., Proc.*, 39: 356—361.

Springer, M.E., 1958. Desert pavement and vesicular layer of some soils of the desert of the Lahontan Basin, Nevada. *Soil Sci. Soc. Am., Proc.*, 22: 63—66.

Stevens, J.H., 1974. Stabilization of aeolian sands in Saudi Arabia's Al Hasa Oasis. *J. Soil Water Conserv.*, 29: 129—133.

Stuart, D.M. and Dixon, R.M., 1973. Water movement and caliche formation in layered arid and semiarid soils. *Soil Sci. Soc. Am., Proc.*, 37: 323—324.

Wischmeier, W.H. and Mannering, J.V., 1969. Relation of soil properties to its erodibility. *Soil Sci. Soc. Am., Proc.*, 33: 131—137.

Wischmeier, W.H. and Smith, D.D., 1965. *Predicting Rainfall-Erosion Losses from Cropland East of the Rocky Mountains. U.S. Department of Agriculture, Agriculture Handbook No. 282.* U.S. Government Printing Office, Washington, D.C.

Woodruff, N.P., Lyles, L., Siddoway, F.H. and Fryrear, D.W., 1972. How to Control Wind Erosion. U.S. Government Printing Office, Washington, D.C., Agriculture Information Bulletin No. 354.

Chapter 12

BIOLOGICAL PROPERTIES

INTRODUCTION

Soil biological activity ranges from a high level at the humid fringe of the arid region to nearly zero in the most arid sections. Soils of the driest part of the Sahara, the dry valleys of the Antarctic, and the center of the Atacama Desert of Chile may be nearly abiotic because of the inhospitality of the climate. In fact, the startling report that no bacteria were found to a depth of 1 m in a soil of the dry Taylor Valley in Antarctica led Horowitz et al. (1972) to make an exhaustive study of a Victoria Valley subsoil in an attempt to determine whether a sterile soil — never before reported on earth in nontoxic soils — actually did exist in Antarctica. They concluded that their soil was truly sterile and, significantly, nontoxic to introduced organisms. Lack of water was the limiting factor. The likelihood that a barren, mobile sand dune in the most arid part of the Sahara — probably the least hospitable environment in the hot deserts — would be sterile appears to be remote since wind-borne dust can bring in bacteria and other soil microorganisms from more hospitable areas. Killian and Fehér (1936) reported a wide variety of microflora in Saharan soils.

Microbial studies of uncultivated soils in arid regions are limited; attention has been centered upon croplands. Consequently, data are scarce on the kinds and numbers of microorganisms present and their distribution, although some progress was made in Desert Biome studies conducted under the International Biological Programme. Cameron (1969) has summarized his extensive investigation of the microorganisms of hot and cold deserts in a diagram (Fig.12.1) of the diversity of flora as influenced by environmental conditions. As the diversity increases, numbers also increase, with bacteria being the most numerous in favorable environments. The dominant microflora include *Bacillus*, *Schizothrix*, *Streptomyces*, *Penicillium*, and *Aspergillus*. As a first approximation, it seems likely that nearly all of the major microbial genera will be present, to a greater or lesser degree, in hot-desert soils if the annual precipitation is 100 mm or more.

ALGAE

Soil algae are among the most important microorganisms in arid-region soils because of their drought resistance, growth habit, and effect on soil

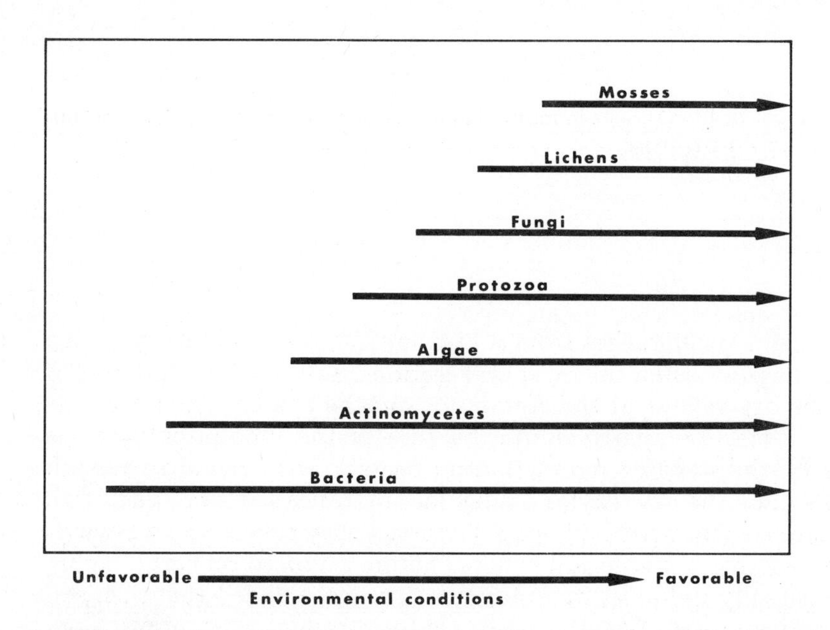

Fig.12.1. Influence of environmental conditions on diversity of microorganisms (adapted from Cameron, 1969).

organic matter and nitrogen content. They grow best in soils having a neutral reaction and they frequently dominate the surface microflora in dry regions. Forming mats on the soil surface as they do, algae serve to bind soil particles together and to shed water; the first effect is an advantage, the second becomes a disadvantage when it increases runoff. Some blue-green algae have the ability to fix atmospheric nitrogen, and their contribution to the nitrogen status of soils is significant in the absence of fertilizer applications.

Nitrogen and organic matter

An incubation study of three Arizona algal soil crusts by Cameron and Fuller (1960) showed that nitrogen content of the crusts increased by 25—42% in four weeks (Table 12.1). The Gila sandy loam crust contained lichens as well as algae and had twice as many genera of algae (16) as the other two. Azotobacter were absent from all three crusts. In pure cultures, nitrogen-fixing algae fixed much more nitrogen than did the crusts of Table 12.1, which were mixtures of algae, soil, and other matter. One calculation of the nitrogen-supplying power of blue-green algae indicated that a soil surface covered by actively growing algae would be capable of fixing 800 kg of nitrogen per hectare per month. In the field, nitrogen fixation rates are much lower because, among other things, only a fraction of the surface has an algal cover. MacGregor and Johnson (1971) in Arizona measured an increase of

TABLE 12.1

Nitrogen fixation by algal soil crusts in moist chambers (from Cameron and Fuller, 1960)

Soil	Incubation time (weeks)	Nitrogen (%)
Tucson sandy loam	0	0.081
	4	0.107
Gila fine sandy loam	0	0.095
	4	0.119
Gila sandy loam	0	0.096
	4	0.136

3—4 g of nitrogen per hour per hectare following a rain. The algal crust covered about 4% of the land area.

Algae add organic matter to soils along with nitrogen, and leaching of organic compounds and nitrates increases the amount of those substances in subsurface layers. Carbon/nitrogen ratios of 14 crusts in the Cameron and Fuller (1960) study ranged from 5 to 18, with an average of 8.5. Nitrate nitrogen was detected in nearly all surface and subsurface soil samples. Annual precipitation is about 280 mm in the study area.

Cameron (1963, 1964) identified a rather bewildering variety of algae in and around the Sonoran Desert in Arizona where the rainfall average ranged from 75 to more than 300 mm. Blue-green algae were represented by 7 families, 32 genera, and 86 species and were by far the most numerous. Other algae (mostly green) belonged to 27 families, 47 genera, and 66 species. Some algae are nitrogen fixers, others are not.

Algal growth

The principal factors controlling algal growth are moisture and sunlight. After a rain, renewed growth of algae begins in less than 30 minutes, under favorable conditions. By far the greatest algal concentration is in the surface few centimeters of soil in small surface depressions where water accumulates after rains. The undersides of partially embedded wood and stones and of translucent rocks also are favored habitats because sunlight is present yet water loss is less rapid than it is from more exposed surfaces. Blue-green algae growth is profuse in irrigated fields of hot arid regions after a rain or irrigation if the vegetative cover is fairly dense. The combination of adequate sunlight, high air temperatures, and reduced evaporation from the shaded soil promote rapid growth of algae.

Fine-textured takyr soils of closed basins have a profuse cover of algae and lichens — but no vascular plants — following the few periods when runoff water collects after rains. However, Lobova (1960) believes that the effect of algae and lichens on takyr formation and humus content is small, contrary to the views of some other Soviet scientists.

Water relations

Microbial crusts have been said to reduce water infiltration and increase runoff in arid regions. Fletcher and Martin (1948) concluded that the opposite was true. They noted that the crusts curled when dry, allowing water to infiltrate while at the same time protecting the soil from the direct impact of raindrops. An additional beneficial effect of the curled crusts is to capture grass seeds when the soil is dry. Upon subsequent wetting, the crusts uncurl and cover the wetted seed, thereby helping germination by reducing evaporation. Microbial crusts had organic carbon and nitrogen contents as much as 300% (for carbon) and 400% (for nitrogen) higher than the underlying soil. The commonly observed phenomenon of better growth of grass seedlings when they are surrounded by microbial crusts indicates that a more favorable plant environment exists where algal, mold, and lichen crusts are present.

AZOTOBACTER

Nitrogen fixation

Nitrogen fixation by free-living (nonsymbiotic) microorganisms is a property shared by both algae and *Azotobacter*. *Rhizobia*, by contrast, are symbiotic nitrogen fixers inhabiting the roots of cultivated and native leguminous plants. There is evidence that some nitrogen fixing organisms have a symbiotic relation with nonleguminous plants.

Azotobacter are responsive to pH conditions: the most favorable pH for nitrogen fixation is that from 6 to 8, which includes the majority of arid-region soils. Moisture is the limiting environmental factor. Sandy soils have fewer azotobacter than finer-textured soils, judging from the data of Peterson and Goodding (1941). These authors speculated that most of the soils of Nebraska (semiarid to subhumid climate) have been inoculated with azotobacter during major dust storms but that survival has depended upon local environmental conditions. Azotobacter activity frequently is limited by the availability of organic carbon for energy.

Growth

In Arizona, Martin (1940) concluded that uncultivated soils in areas having an annual rainfall between 75 and 375 mm were generally lacking in azotobacter. Only about 23% of the soils contained any of the organism and they were usually inactive in fixing nitrogen. Eighty seven percent of the irrigated soils in the same rainfall zones contained azotobacter, and they were generally active. The scarcity and inactivity of azotobacter in the uncultivated soils appeared to be the result of high temperatures, low humidity, dry soil for most of the year, and low soil organic-matter content. Similar results were obtained by Vandecaveye and Moodie (1942) in climatic zones of central Washington receiving 150—375 mm of precipitation. They concluded that azotobacter populations in irrigated and nonirrigated soils of the region probably are too small to be a significant factor in nitrogen fixation.

SOIL AGGREGATION

Aggregate formation

Microorganisms play an important, if not dominant, role in the formation and stability of soil aggregates. Secretion of microbial gums was long thought to be the principal mechanisms by which individual soil grains were bound together. Significant though that effect is, Hubbell and Chapman (1946) concluded that it is not the main cause of aggregate stability. Their experiments with two calcareous soils showed that aggregates formed only in the presence of living organisms. Bacteria were capable of forming small and fragile water-stable aggregates which were easily crushed, whereas larger aggregates having the filaments of fungi and actinomycetes in and around them were resistant to crushing and were persistent. Microbial secretions, in the absence of living organisms, were incapable of forming water-stable aggregates. Roots contributed to aggregate formation only after it had been initiated microbially. Bond and Harris (1964) confirmed the results for fungi and found that algae have a similar effect. These authors also noted that water repellence of dry sands was associated with the growth of microorganisms.

Erosion control

In arid regions of Australia, many plant communities have a relatively open canopy exposing bare areas of erodible sand. Where the vegetation consists of low shrubs, crusts dominated by algae in association with bacteria and fungi promote aggregate formation and are good soil stabilizers (Bond and Harris, 1964). The crusts usually are about 3 mm thick, and sometimes twice that thick. Undisturbed crusts effectively resisted sand movement by wind and water but once the crust was broken by animals, wind scouring occurred.

ORGANIC MATTER

Equilibrium levels

Average organic-matter levels in arid-region soils generally range from less than 0.5% in the drier areas to 2—4% in the semiarid zone. Equilibrium levels are dependent upon the rate of organic-matter additions by roots and plant debris and the rate of decomposition by microorganisms. Each climatic zone has a characteristic equilibrium level, under natural conditions, where plant growth and decomposition over a period of years become equal. When irrigation is introduced, soil organic-matter content usually will increase as a result of the increased plant growth and a greater return of plant residues to the soil. Dryland farming, on the other hand, generally leads to lower organic-matter levels than were present in the original soil.

Textural effects

Among the factors controlling organic-matter levels, soil texture is of major importance. Within any one climatic zone, all other conditions being equal, fine-textured soils will contain more organic matter than coarse-textured soils. Data from Dregne and Maker (1955) demonstrated the effect of soil texture on organic-matter content in an irrigated area of southern New Mexico (Table 12.2). The absolute amount of organic matter would be less in nonirrigated soils in the same climatic zone but the textural relations would be similar. Organic matter adsorbed on soil particles is more difficultly decomposable than when it is loose in the soil. Since fine-textured soils have a much greater surface area and are, therefore, capable of absorbing more

TABLE 12.2

Percentage of organic matter in irrigated soils having different textures (from Dregne and Maker, 1955)

Soil texture	Organic matter (%)
Clay, silty clay	1.8
Clay loam, silty clay loam, sandy clay	2.0
Loam, sandy clay loam, silt loam	1.6
Sandy loam	1.0
Loamy sand, sand	0.3

organic matter than coarse-textured soils, the rate of decomposition is less in clays than in sands. Temperature and moisture differences also play a part.

Distribution in shrublands

In heavily grazed desert grasslands where shrubs are present, soil organic-matter content commonly is higher under the shrubs than between them. Tiedemann and Klemmedson (1973) found that not only was the surface organic-matter content three times greater under mesquite (*Prosopis juliflora*) trees than between the mesquite, the availability of soil nitrogen to native perennial grasses was up to 15 times higher under the trees. Phosphorus availability also was significantly higher under the mesquite. Low nutrient availability in the open areas is likely, then, to be responsible for at least part of the forage-production problem in desert grasslands.

Carbon/nitrogen ratios

Carbon/nitrogen ratios of arid-region soils, on the average, increase from the more arid to the less arid areas. Ratios as low as 6 or 7 have been reported for surface soils in rainfall zones of less than 150 mm where vascular plants are absent. As precipitation increases, the ratio rises to 10 or 11 in the wet end of the semiarid zone. Low C/N ratios indicate a greater degree of decomposition of organic residues than do high ratios. Hot regions tend to have lower ratios than cold regions.

PLANT—SOIL RELATIONS

Plants as soil indicators

The fact that plants modify the soil environment in which they are growing and also are affected by soil conditions has long been known. In the arid regions, attempts have been made to relate the presence or absence — or the general state of health — of single shrub species to soil characteristics. Thus, the indicator significance of *Prosopis*, *Larrea*, *Atriplex*, *Artemisia*, and other shrubs for the selection of land suitable for irrigation has been investigated, with varying degrees of success. Current thinking is that the composition of the plant community is more significant than the status of a single species in evaluating soil conditions. Soil-vegetation maps have demonstrated the relation of plant communities to soil types in the arid regions, particularly with respect to soil physical properties (texture, depth to rock, etc.).

Despite the importance of communities, individual shrubs do have a profound effect on the soil under their canopy. The data of Tiedemann and Klemmedson (1973) are just one example of the notable accumulation of organic matter and soluble salts under *Prosopis* as compared to soils outside the shrub canopy. A classical study by Fireman and Hayward (1952) of soil—plant relations in the Escalante Desert of Utah pointed out that *Artemisia tridentata* had no detectable effect, whereas *Sarcobatus vermiculatus* had a major effect on soil salinity, pH, and exchangeable-sodium percentage.

Toxic exudates

Some shrubs are said to release an exudate or to contain within their leaves a toxin which inhibits germination of other seeds and accounts for the poor growth of competing plants in the vicinity. A desert grassland study of the effect of *Larrea divaricata* (creosotebush) plant parts on seed germination showed that water extracts from all parts of the plant (stems, leaves, roots) inhibited germination of some, but not all, grasses (Knipe and Herbel, 1966). Furthermore, the carpels of the shrub contain a substance that inhibits germination of the creosotebush seed, itself.

REFERENCES

Bond, R.D. and Harris, J.R., 1964. The influence of the microflora on physical properties of soils. I. Effects associated with filamentous algae and fungi. *Aust. Soil Res.*, 2: 111—123.

Cameron, R.E., 1963. Algae of southern Arizona. Part I. Introduction — blue-green algae. *Rev. Agrol.*, 7: 282—318.

Cameron, R.E., 1964. Algae of southern Arizona. Part II. Algal flora (exclusive of blue-green algae). *Rev. Agrol.*, 7: 151—177.

Cameron, R.E., 1969. Abundance of microflora in soils of desert regions. Jet Propulsion Laboratory, California Institute of Technology, Pasadena, Calif., Technical Report 32-1378.

Cameron, R.E. and Fuller, W.H., 1960. Nitrogen fixation by some algae in Arizona soils. *Soil Sci. Soc. Am., Proc.*, 24: 353—356.

Dregne, H.E. and Maker, H.J., 1955. Fertility levels of New Mexico soils. *N.M. Agric. Exp. Stn., Bull.*, No. 396.

Fireman, M. and Hayward, H.E., 1952. Indicator significance of some shrubs in the Escalante Desert, Utah. *Bot. Gaz.*, 114: 143—155.

Fletcher, J.E. and Martin, W.P., 1948. Some effects of algae and molds in the rain crust of desert soils. *Ecology*, 29: 95—100.

Horowitz, N.H., Cameron, R.E. and Hubbard, J.S., 1972. Microbiology of the dry valleys of Antarctica. *Science*, 176: 242—245.

Hubbell, D.S. and Chapman, J.E., 1946. The genesis of structure in two calcareous soils. *Soil Sci.*, 62: 271—281.

Killian, C. and Fehér, D., 1936. La fertilité des sols du Sahara. *Genie Rural*, 108(2790): 114—124.

Knipe, D. and Herbel, C.H., 1966. Germination and growth of some semidesert grassland species treated with aqueous extract from creosotebush. *Ecology*, 47: 775—781.

Lobova, E.V., 1960. *Pochvy Pustynnoi Zony SSSR.* (*Soils of the Desert Zone of the U.S.S.R.* Issued in translation by the Israel Program for Scientific Translations, Jerusalem, 1967, 405 pp.; also cited as TT 67-61279.)

MacGregor, A.N. and Johnson, D.E., 1971. Capacity of desert algal crusts to fix atmospheric nitrogen. *Soil Sci. Soc. Am., Proc.*, 35: 843—844.

Martin, W.P., 1940. Distribution and activity of azotobacter in the range and cultivated soils of Arizona. *Ariz. Agric. Exp. Stn., Tech. Bull.*, No. 83.

Peterson, H.B. and Goodding, T.H., 1941. The geographic distribution of *Azotobacter* and *Rhizobium meliloti* in Nebraska soils in relation to environmental factors. *Nebr. Agric. Exp. Stn., Res. Bull.*, No. 121.

Tiedemann, A.R. and Klemmedson, J.O., 1973. Nutrient availability in desert grassland soils under mesquite (*Prosopis juliflora*) trees and adjacent open areas. *Soil Sci. Soc. Am., Proc.*, 37: 107—111.

Vandecaveye, S.C. and Moodie, C.D., 1942. Occurrence and activity of *Azotobacter* in semiarid soils in Washington. *Soil Sci. Soc. Am., Proc.*, 7: 229—236.

GLOSSARY

Base saturation percentage. The percentage of the cation-exchange capacity occupied by basic cations. Calcium, magnesium, sodium, and potassium ions are the most common basic cations. Hydrogen and aluminum ions are acidic cations.

Calcareous soil. A soil containing sufficient carbonates to effervesce visibly when treated with cold dilute hydrochloric acid. Calcium carbonate and magnesium carbonate are the principal carbonates in soils.

Cation-exchange capacity. The amount of exchangeable cations a unit weight of soil is capable of holding. Cation-exchange capacity is expressed as milliequivalents of exchangeable cations per 100 g of soil.

Exchangeable cation. A positively charged ion adsorbed on the surface of a soil particle and capable of being replaced by another cation in the soil solution.

Exchangeable sodium. Adsorbed sodium ions capable of being replaced by other cations in the soil solution.

Lime. Calcium carbonate and magnesium carbonate.

Loess. Deposits of wind-blown material, dominantly composed of silt-size mineral particles. Loess is extensive in Argentina, China, the Soviet Union, and the United States.

Mottling. Spots of color on a contrasting soil background. The most prominent mottling consists of orange and red spots on a uniform brown or gray background color; such mottles indicate that impeded drainage has occurred.

Mulch. A layer of straw, plastic, gravel, or other material spread on the surface of the soil to control evaporation, erosion, heat transfer, etc.

Nitrogen fixation. The conversion of elemental atmospheric nitrogen to combinations of nitrogen with hydrogen or oxygen. Specialized soil microorganisms are responsible for most nitrogen fixation in nature.

Soil depth. Depth of effective soil material over bedrock, shale, gypsiferous earth, petrocalcic layers, or other materials in which plant-root development is restricted. Depth classes used in this book are:

Class	*Depth over ineffective material (cm)*
very shallow	<25
shallow	25—50
moderately deep	50—100
deep	100

Soil permeability. The rate at which air or water move through undisturbed soil after equilibrium has been established. Water-permeability rates are grouped as follows:

Group	*Water-permeability rate (cm/hr)*
very slow	<0.5
slow	0.5—1.5
moderate	1.5—5
rapid	5—15
very rapid	>15

Soil structure. The arrangement of primary soil particles (sand, silt, clay) into secondary particles (peds, blocks, columns, prisms, etc.).

Soil texture. The relative proportions of sand, silt, and clay in a mass of soil. Particle sizes are:

Particle	*Size (diameter in mm)*
clay	<0.002
silt	0.002—0.05
sand	0.05—2.0

Gravel is 2—75 mm in diameter. Soils containing more than 15% gravel have the adjective "gravelly" placed before the textural class name (e.g., gravelly sandy loam).

INDEX

*Italicized pagenumbers refer to illustrations.